Safety and reliability of existing structures

Safety and reliability of existing structures

James T. P. Yao
Purdue University

Pitman Advanced Publishing Program
Boston · London · Melbourne

PITMAN PUBLISHING INC
1020 Plain Street, Marshfield, Massachusetts 02050

PITMAN PUBLISHING LIMITED
128 Long Acre, London WC2E 9AN

Associated Companies
Pitman Publishing Pty Ltd, Melbourne
Pitman Publishing New Zealand Ltd, Wellington
Copp Clark Pitman, Toronto

First published in Great Britain 1985

Library of Congress Cataloging in Publication Data
Yao, James Tsu-ping, 1932–
Safety and reliability of existing structures.
(Surveys in structural engineering
and structural mechanics)
Bibliography: p.
Includes index.
1. Structural stability. 2. Structural failures.
I. Title. II. Series.
TA656.Y36 1984 624.1′028′9 84-994
ISBN 0-273-08582-4

British Library Cataloguing in Publication Data
Yao, James T.P.
Safety and reliability of existing structures.
—(Surveys in structural engineering
and structural mechanics; 2)
1. Structures, Theory of. 2. Safety factor
in engineering
I. Title. II. Series.
624.1′71 TA656.5
ISBN 0-273-08582-4

Typeset and printed in Great Britain at The Pitman Press, Bath

Contents

Preface

Although many practising engineers have been successfully evaluating the safety of existing structures throughout the history of structural engineering, much of the decision-making process has depended on each engineer's experience, intuition, and judgment. In this book, I attempt to examine this problem from several different viewpoints. Several damage functions for civil engineering structures are summarized and reviewed along with the current practice. Various system identification techniques in structural dynamics are also discussed in terms of their potential applications for the evaluation of structural safety. To help understand how experts summarize and interpret results of measurements, inspection, and analyses in reaching their decision concerning structural safety, the application of rule-inference methods is reviewed and discussed. A direct link between such methods and the classical theory of structural reliability is also suggested herein.

As a life-long student of structural safety and an enthusiastic novice of the failure behavior of existing structures, I am most interested in learning all aspects of this challenging problem. I am most fortunate in having many good friends who are experts in related subject areas both in academic institutions and in private practice. Moreover, I am indebted to my co-workers including B. Bresler, C. B. Brown, S. J. Hong Chen, K. S. Fu, T. V. Galambos, G. C. Hart, J. M. Hanson, M. Ishizuka, S. Toussi for their valuable collaboration. The support of National Science Foundation and encouragement of M. P. Gaus, S. C. Liu, J. Scolzi, and J. E. Goldberg are gratefully acknowledged. The Main Editor of the Monograph Series, W. F. Chen, and Associate Director, John Hindley, of Pitman Publishing Limited provided the necessary impetus and prodding, without which this work would not have been completed. Marian Sipes capably typed the first draft of the manuscript.

J. T. P. Yao

1 Introduction

1.1 General remarks

Various activities in the structural engineering profession have been summarized by Galambos and Yao (62) in terms of the state of nature (the way things are) and the state of art (the body of knowledge). In the state of nature, there exist human and societal needs, environmental conditions, man-made structures, response of these structures to environmental conditions and their consequence and utility. The primary objective of structural engineers is to design these structures to obtain specific structural behavior with desirable consequences and thus satisfy their intended functions to meet certain human and societal needs. On the other hand, the state of the art is an idealization of the complex phenomena in the state of nature for the purpose of making structural analysis and design. Such idealized and mathematical models are in need of continuous updating and improvement.

The interrelationship between the state of nature and the state of art of structural engineering is schematically shown in Fig. 1. In the state of nature, a structure is subjected to excitation (or disturbance or load) such as winds and earthquakes throughout its intended lifetime. The structural response to such excitation can be found in the form of displacements, internal forces, stresses and strains, etc., which present a 'demand' on the structure. Inherently, each structure possesses a 'capacity', which consists of various limit states. Damage or failure may result whenever the demand exceeds the capacity (or one or more components in the response exceed the corresponding limit state(s)).

In general, the structural system and its environment are idealized so that mathematical analyses can be made. The characteristics of a structure can usually be modeled with simple equations. For example, the maximum moment m of a simply-supported beam with length l and a uniformly distributed load w can be found as $wl^2/8$. As another example, the motion of mass–damper–spring systems can be represented with a set of ordinary differential equations. Whenever the load (or disturbance) and the structural model are given or assumed, the process to find the response is called

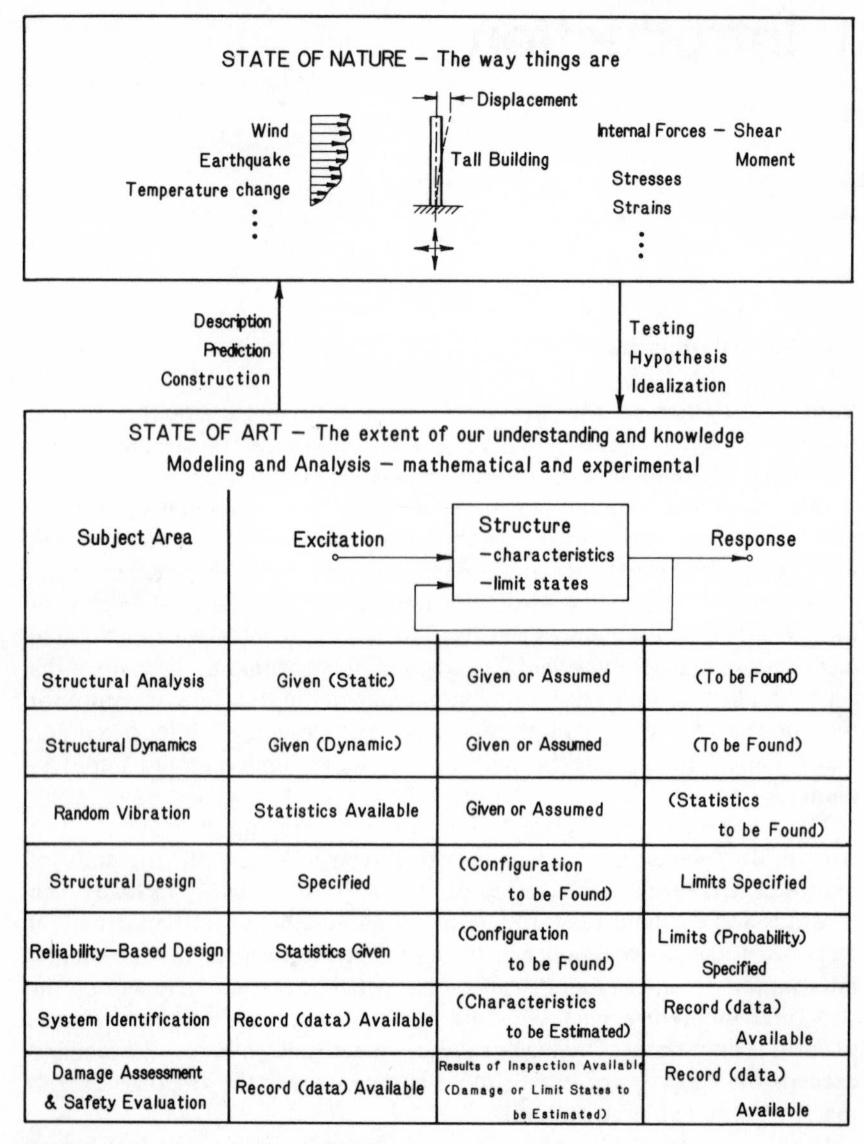

Subject Area	Excitation	Structure	Response
Structural Analysis	Given (Static)	Given or Assumed	(To be Found)
Structural Dynamics	Given (Dynamic)	Given or Assumed	(To be Found)
Random Vibration	Statistics Available	Given or Assumed	(Statistics to be Found)
Structural Design	Specified	(Configuration to be Found)	Limits Specified
Reliability–Based Design	Statistics Given	(Configuration to be Found)	Limits (Probability) Specified
System Identification	Record (data) Available	(Characteristics to be Estimated)	Record (data) Available
Damage Assessment & Safety Evaluation	Record (data) Available	Results of Inspection Available (Damage or Limit States to be Estimated)	Record (data) Available

Fig. 1 Subject areas in structural engineering

structural analysis. More specifically, when the load is given as a function of time, the subject area becomes known as structural dynamics (100). When the dynamic excitation is random and its certain statistics are available, the methodology for finding essential statistics of the structural response is called random vibration (35). These response statistics can then be used along with resistance (limit states) statistics for the estimation of structural reliability (89).

On the other hand, it is necessary to determine the configuration (including geometry of the structure and size of its members), the material, and the type of construction before the structure is built. In the design process, the loading conditions and limits of the response are specified and the dimensions of the structure are usually sought. As it is well known, the design of a structure frequently follows an iterative process involving both structural analysis and structural design. Because most techniques of structural analysis are applicable only to idealized and simplified systems, the behavior of a completed structure in the state of nature may not correspond to that of the original mathematical model. For certain important structures, nondestructive tests are performed with selected load and response data collected. Techniques of system identification are then applied to obtain a more realistic model for further analysis.

Since the late Professor A. M. Freudenthal presented a rational approach to the structural safety problem more than thirty-five years ago (53), an ever-increasing effort has been directed toward the application of the theory of probability and statistics in structural engineering. In the classical theory (4, 54), the probability of failure or survival for a given type of structure is computed with assumed distributions of random variables representing the load and the corresponding resistance. For structures subjected to dynamic loads, random processes are applied (e.g., (35, 89)). In recent years, several reliability-based design specifications have been developed and adopted around the world (e.g., (2, 5, 34, 104)).

For structures which have been constructed and are thus existing, it is desirable and frequently necessary to assess their respective damage states on the basis of available information including measured and recorded experimental data. In addition, it is desirable to re-evaluate the reliability calculations of these structures so that rational decisions can be made in regard to any necessary repairs, replacements, retirement, and other maintenance or rehabilitation processes.

Whenever it is necessary, a structure can be designed to satisfy code requirements and to perform satisfactorily on the basis of past experience and available knowledge. A site is selected and field data are collected. Usually, a preliminary design is made and the idealized mathematical representation is analyzed for expected or specified loading conditions. Based on these analytical studies, the design may be revised and re-analyzed in an iterative manner until all design criteria are satisfied. The completed design is then implemented through construction as shown schematically in Fig. 2.

Because (*a*) it is difficult to predict future loading conditions and (*b*) materials in the structure possess random characteristics, random

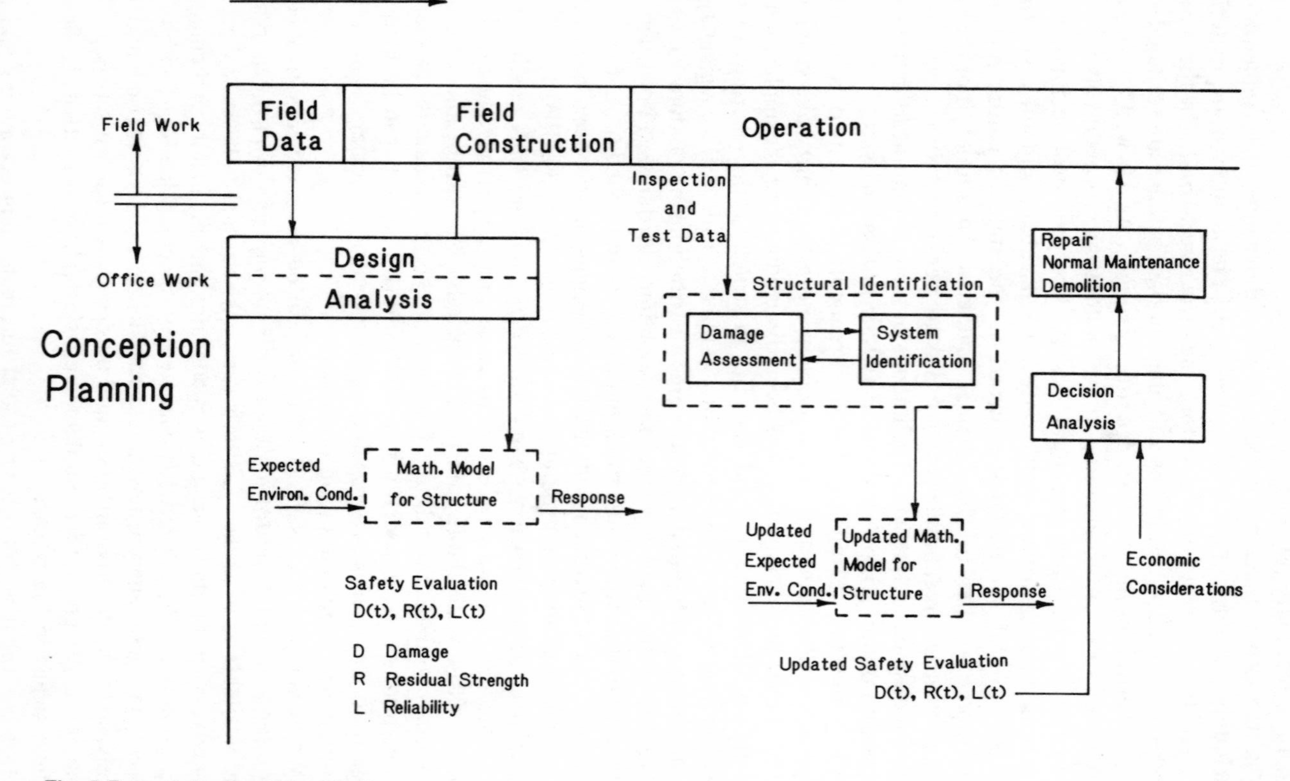

Fig. 2 Role of structural identification

processes have been used to represent these quantities for estimation of failure probabilities. However, the as-built structure is usually different from the original mathematical model in the design process. The fact is that the real-world structure is an extremely complex system. Even with the use of finite element methods and modern computers, it is usually impractical and unfeasible to consider all the details in the mathematical model of a given structure. Moreover, the damage path and failure behavior of most large structures remain unknown because few experimental studies of full-scale structures are available.

For certain important structures, nondestructive dynamic tests are conducted for the estimation of dynamic properties of the as-built structure. These test data are then used to obtain 'improved' or 'more realistic' equations of motion. These equations of motion are applicable within the range of the test amplitude, which is usually small and within the linear behavior of the given structure. Therefore, one cannot apply results of such analysis to include the consideration of destructive or damaging loading conditions. Nevertheless, these mathematical representations can be useful for comparison purposes. For example, any change in the measured natural frequencies may be used as an indicator of structural damage.

Generally speaking, a select group of experienced structural engineers can investigate the condition of a particular structure and determine its level of safety. In such investigations, the original design calculations and drawings (if available) are examined and checked. Inspections and testings are conducted, and the resulting data are analyzed. Results of these analyses are then summarized and interpreted by experienced engineers to yield appropriate recommendations. Although it is possible to understand the inspection and testing conducted and the detailed analysis performed in such studies, the decision-making process involved in deciding (a) specific types of inspection and testing procedures and (b) the summary and interpretation of experimental and analytical results remain the privileged information of relatively few experts in the structural engineering profession.

1.2 Objective and scope

The objective of this book is to summarize and discuss the state of the art of several subject areas related to safety and reliability of existing structures. In Chapter 2, the available damage functions for civil engineering structures are reviewed and discussed. In Chapter 3, current practices for the safety evaluation of existing structures are summarized. In Chapter 4, the role of system identification and its potential application to structural dynamics and reliability studies are explored. In Chapter 5, the possible application of rule-inference methods to the assessment of damage is introduced and examined. An attempt is also made in Chapter 6 to propose a unified approach for the solution of problems concerning safety and reliability of existing structures.

2 Damage functions for civil engineering structures

2.1 General remarks

It is desirable to find a generally acceptable and meaningful definition of structural damage for various types of structures. With such a measure of structural damage, the safety of a specific existing structure can be assessed and appropriate decisions regarding repair and maintenance can be made accordingly. In this book, the term 'damage' refers to any deficiency and/or deterioration of strength as caused by external loading and environmental conditions as well as human errors in design and construction. Therefore, a poorly designed and/or poorly constructed structure can have an initial 'damage' measure while it is still new without experiencing any severe loading conditions.

Generally, there are three types of definitions for structural damage. The first one is numerical, the second one is given in terms of repair or replacement costs, and the third one is verbal. Frequently, numerical values are also assigned to various verbal classifications. Some of these definitions are reviewed in Section 2.2. Several problems remain in applying these damage functions in engineering practice. These problems and their possible solutions are presented in Section 2.3. A general discussion of these topics is given in Section 2.4.

2.2 Available damage functions

In 1971, Wiggins and Moran (133) developed a procedure for grading existing buildings in Long Beach, California. A total of up to 180 points is assigned to each structure according to the evaluation of the following five items:

(*a*) *Framing system and/or walls* (0, 20, 40 points) A well-designed reinforced concrete or steel building less than three stories in height is assigned a zero-value. On the other hand, an unreinforced masonry filler and bearing walls with poor quality mortar is assigned a value of 40 points.

(*b*) *Diaphragm and/or bracing system* (0, 10, 20 points) As an example, zero values correspond to well-anchored reinforced slabs and fills. On the other hand, incomplete or inadequate bracing systems correspond to the high 20 points on the scale.

(*c*) *Partitions* (0, 10, 20 points) Those partitions with many wood or metal stud bearings rate zero points. On the other hand, unreinforced masonry partitions with poor mortar will draw 20 points.

(*d*) *Special hazards* (0, 5, 10, 15, 20, 35, 50 points) The high hazards include the presence of non-bearing, unreinforced masonry walls, parapet walls, or appendages.

(*e*) *Physical condition* (0, 10, 15, 20, 35, 50 points) The high hazards include serious bowing or leaning, signs of incipient structural failure, serious deterioration of structural materials, and other serious unrepaired earthquake damage.

For each building thus inspected, all these five numbers are added. The sum may be considered as a damage index. Rehabilitation is not required if the sum is less than 50 points (low hazard). Some strengthening is required if the sum is between 51 and 100 points (intermediate hazard). Demolition or major strengthening is necessary when the sum exceeds 100 points (high hazard).

Detailed guidelines are given for the assignment of numbers in each category. Therefore, this method is relatively simple to use even for inspectors who are not trained as engineers. However, it is difficult to develop such a simple procedure to include all special cases. Moreover, the demarcation between low, intermediate, and high hazards is rather arbitrary for these verbal terms which cannot be clearly defined.

In 1975, Culver *et al.* (36) proposed the field evaluation method (FEM) which is applicable even when building plans are unavailable. A rating of 1 through 4 is assigned for each of the following factors: (*a*) general rating, GR, for grading the materials of the frame; (*b*) structural system rating, s, for combining ratings of connections, roofs, and floors, etc.; and (*c*) Modified Mercalli Intensity I. Then a composite rating, CR, is computed as follows:

$$CR = \frac{GR + 2s}{eI} \tag{1}$$

If $CR < 1.0$; the building is said to be in good condition, if $1.0 \leqslant CR \leqslant 1.4$; it is in fair condition, if $1.4 \leqslant CR \leqslant 2.0$; it is in poor condition, if $CR > 2.0$; it is in very poor condition.

In addition, a more detailed methodology was also presented for survey and evaluation of existing buildings to determine the risk to life safety under natural hazard conditions and estimate the amount of expected damage. There are four major parts in this report as follows:

(*a*) generation of site loads,
(*b*) generation of a structural model,
(*c*) computation of response, drift and ductility, and
(*d*) assessment of damage.

The damage on *i*th story, D_i, resulting from extreme natural environments is expressed in terms of total damage as follows:

$$D_i = F\left(\frac{\Delta_i}{(\Delta_y)_i}\right) \tag{2}$$

where $F(\cdot)$ = distribution function
Δ_i = calculated interstory drift of *i*th story
$(\Delta_y)_i$ = user specified interstory drift corresponding to yielding of *i*th story.

The damage is classified into three categories: structural, nonstructural and glass. It is further subdivided into frame, walls and diaphragms in the case of structural damage.

More generally, the damage can be expressed as sum of the 'initial' damage D_0, and a function of the ductility. Usually, the initial damage is assumed to be zero. Blume and Monroe (13) assumed that the structural damage is linearly related to ductility factor with '0' denoting elastic behavior and '1' denoting collapse. Bertero and Bresler (11) stated that (*a*) the lateral displacement ductility factors generally provide a good indication of structural damage, and (*b*) the interstory drift is a more important factor in causing nonstructural damage. The interstory drift of reinforced concrete buildings during earthquakes was studied extensively by Sozen (1, 99, 120). Bresler (15, 16) discussed the relative merits of using plasticity ratio (residual deformation to yield deformation) and the ductility. For structures which are subjected to cyclic plastic deformations with decreasing resistance, the ratio of the initial to *j*th-cycle resistance at the same cyclic peak deformation was also suggested.

For monotonic loading conditions, Oliveira (102) defined a damage ratio function, D, as follows:

$$D = \left\langle\frac{Z-y}{c-y}\right\rangle^b \tag{3}$$

where Z = maximum displacement response
y = yield displacement
c = displacement at collapse
b = material and structural parameter
$\langle x\rangle^n$ = singularity function such that, for $n \geqslant 0$, $\langle x\rangle^n = 0$ when $x < 0$, and $\langle x\rangle^n = x^n$ when $x \geqslant 0$.

For axially-loaded mild steel specimens which are subjected to low-cycle high-amplitude reversed plastic deformations, Yao and Munse (143) suggested the use of the following damage function:

$$D = \sum_{i=1}^{n} \left[\left(\frac{\Delta q}{\Delta q_1} \right)^{1/m} \right]_i \tag{4}$$

where $1/m$ = a parameter depending upon the ratio of the cyclic compressive change in plastic strain to the subsequent tensile change in plastic strain,
Δq = percent cyclic tensile change in plastic true strain,
Δq_1 = percent cyclic tensile change in plastic true strain at $n = 1$,
n = number of applications of tensile load prior to fracture.

It is interesting to note that Eq. 3 may be considered as a special case of Eq. 4 with the following correspondence:

$$\Delta q = Z - y$$
$$\Delta q_1 = C - y$$
$$1/m = b$$

Lacking for a well-established cumulative damage law for structural systems, Eq. 4 was applied to evaluate the damageability of seismic structures by Kasiraj and Yao (82, 83) for a given earthquake, and later by Tang and Yao for random ground motions (123, 124). Recently, Rosenblueth and Yao (107) used the following damage function in their study of cumulative damage of seismic structures:

$$D = \sum_{i=1}^{n} a_i \left\langle \frac{Z}{y} - 1 \right\rangle^{b_i} \tag{5}$$

where a_i and b_i are empirical constants. Although more full-scale and destructive tests are conducted in recent years (7, 61, 72, 95), currently available test data are still insufficient to either validate the form of such a damage function or to estimate these parameters for reinforced concrete structures.

Aristizabal-Ochoa and Sozen (6) used a damage ratio, which is comparable to but not exactly the same as the ductility. The damage ratio is used in the substitute-structures method, with which the inelastic response of the structure can be considered by using a linear dynamic analysis. Okada and Bresler (19, 101) discussed a screening method, in which the reinforced concrete buildings are classified according to three types of failure mechanisms (bending, shear and shear-bending) by considering nonlinear response of the structure to two levels of earthquake motion (0.3 g and

$0.45g$). The 'first screening' deals with approximate evaluation of the load–deflection characteristics of the first story or of the weakest story. The 'second screening' consists of a time-history nonlinear response analysis of each story. The 'third screening' makes use of a dynamic response analysis including the nonlinearity of each member.

Bertero and Bresler (11) presented damageability criteria according to local, global, and cumulative damage using the summation operation. An importance factor is introduced for each element depending upon such considerations as life hazard and cost. The resulting normalized cumulative damage index is given as follows:

$$D = \frac{1}{\Sigma w_i} \sum_{i=1}^{m} \frac{w_i \eta_i s_i}{\chi_i r_i} \tag{6}$$

where w_i = importance factor for ith structural element,
η_i = service history influence coefficient for demand,
s_i = response (or demand) in the ith element due to load,
χ_i = service history influence coefficient for capacity,
r_i = resistance (or capacity) in the ith element.

The rational determination of such factors as w, η, and χ in Eq. 6 has been studied in the case of a hypothetical three-story reinforced concrete building (12). Nevertheless, it is necessary to conduct further investigations to make this method more practical and useful in engineering practice.

Recently, Bresler and Hanson (17) suggested the following cumulative damage function for a prescribed series of events k:

$$D_k = \frac{\sum_{i=1}^{m} w_{ik} d_{ik}}{\sum_{i=1}^{m} w_{ik}} \tag{7}$$

where

$$d_{ik} = d_{ij} + \left(\frac{Z_{ik} - y_{ik}}{c_{ik} - y_{ik}}\right)(1 - d_{ij}) \tag{8}$$

where w_{ik} = cumulative importance factor for the ith element and events k
d_{ik} = local damage index for ith element and events k
d_{ij} = local damage index for ith element and events $j < k$
Z = demand in terms of displacements or other functions
c = capacity for displacement or other functions
y = threshold for yielding (or other limit states).

They concluded that 'further studies are needed to establish appropriate criteria for values of damage threshold and of ultimate damage capacity for different types of structural elements under different loading conditions, and different loading histories'. The need for the determination of the importance or weighting factors was also stated.

Lee and Collins (88) developed a systematic methodology for the determination of risk for structures due to fire, flood, earthquake, wind hazards. The risk equation was used to obtain an estimate of the average annual loss. In this study, the damage was represented by percent of replacement value of the structure.

Hasselman and Wiggins (70) applied Bayesian statistics to combine earthquake damage data (in lower damage states) with interstory drift (in high damage states). In their models, the type of construction, materials used, the quality of design and construction, and the duration of strong earthquake in relation to the fundamental period of the building are considered. Applications were made to three actual buildings including one building damaged during the 1971 San Fernando Earthquake.

Kustu *et al.* (86) collected laboratory test data in the form of force–deformation relationships. They defined a set of damage states for each type of component. As an example, a reinforced concrete beam can be 'uncracked', 'cracked but unyielded', 'yielded', and 'failed'. Either a normal or a lognormal distribution was used to represent the statistical variations of various damage threshold values. A 'central damage factor', γ, is defined as the ratio of the estimated cost of repairs for a damage state to the replacement value of the component in question. The conditional probability $P(D_i|v)$ of a component being in damage state D_i given a calculated demand v is then calculated. Finally, the expected damage can be calculated as follows:

$$E(D) = \Sigma \gamma P(D_i|v) \tag{9}$$

Such a procedure is logical and the report was well written. Nevertheless, the following questions may be raised. First, there are too many types of components and too many different designs to obtain sufficient data for detailed statistical analysis in all of these cases. Second, the demarcations between adjacent damage states are not clearly defined. For instance, one may have difficulties in distinguishing the 'uncracked' from the 'cracked but unyielded' damage states in actual structures. Moreover, the yielding of a reinforced concrete beam is not as definitive as that in a coupon specimen of mild steel. Third, there are questions concerning the calculation of demand v in the component because the failure behavior of complex structures remains unknown in most cases. To date, there does not seem to exist any adequate mathematical model for such analyses.

Finally, it seems that it is difficult to evaluate the central damage factor in reality.

The Modified Mercalli Intensity (MMI) scale (100) is an earlier descriptive classification of structural damage. In studying the building damage resulting from the Caracas Earthquake of 29 July 1967, Seed *et al.* (112) used several quantities to represent the damage state of buildings. For each individual building, the ratio of maximum induced dynamic lateral force to static design lateral force is used for brittle structures, and the ratio of spectral velocity to lateral force coefficient is used for ductile structures. Shinozuka and Kawakami (115) reported on the use of a 'leakage (or break) damage index' in studying earthquake damage of underground pipeline systems in Japan. This index is given as the ratio of the number of pipe breakages to the length (km) of the pipelines in each area.

In 1973 Whitman *et al.* (134) defined several damage states for use in a damage matrix to evaluate the damageability of various classes of buildings. In an application in estimating structural damage due to tornadoes, Hart (66) gave six classifications as 'none', 'light', 'moderate', 'heavy', 'very severe', and 'collapse' on the basis of the ratio of repair cost to replacement cost for the entire structure. Hsu *et al.* (74, 75) used a similar scale in their study of seismic risks in 1976. Recently, Whitman *et al.* (135) studied two specific buildings in Boston to evaluate their as-built resistance using four categories of damage state, namely, none or minor, slight or moderate, serious, and total damage. Housner and Jennings (73) used classifications such as minor, moderate, severe, major damage, and partial collapse. A similar classification system is recommended in a publication of the Earthquake Engineering Research Institute (87). In Yugoslavia (79), the buildings may be classified as usable (undamaged, and slightly damaged); temporarily unusable (damaged structural system, and considerably damaged structural system); and unusable (heavily damaged structural system, and partial and complete failure).

2.3 Problems and possible solutions

Engineers usually like to manipulate numbers. Therefore, those definitions of structural damage with numerical values are appealing to many engineers. However, it appears that the several numerical scales as summarized in Section 2.2 are rather arbitrary. Those definitions involving repair or replacement costs are attractive on the surface. Nevertheless, questions remain as to the process with which such costs can be determined rationally and accurately. The verbal classifications are meaningful, especially if and when such classifications are made by experts. Nonetheless, there exist cases when numerical values are needed for further analysis of structural damage. As an example, if a particular structural member is severely damaged, how does it affect the damage state of the entire structure? In the following, a possible approach is suggested.

Following Blockley (14) and Brown (20), a safety measure, N, is defined as the negative logarithm (with base 10) of the probability of failure, i.e.,

$$N = -\log_{10} p_f \tag{10}$$

The most important advantage for such a safety measure is that it is directly related to the probability of failure in a meaningful manner. Furthermore, it can be determined both objectively (4, 53, 54) and subjectively (14, 20). As it will be discussed in more detail later, such a safety measure can be related to verbal classifications in a meaningful way.

2.4 Discussion

Several definitions of structural damage are summarized and discussed in this chapter. As stated earlier, a brand new structure may or may not be safe because it may be defective due to errors in analysis and design, fabrication, and/or construction. Consequently, any structure may have an initial value of 'damage' in the broad sense. Certainly, there can be additional damage as results of unexpected usage of the structures, wear and corrosion, the occurrence of hazardous events, fatigue and other cumulative damage.

To illustrate the fact that there are two types of problems in safety evaluation of structures, we use the following classical reliability function (54):

$$L_T(t) = L_T(0) \exp\left[-\int_0^t h_T(t)\mathrm{d}t\right] \tag{11}$$

where $h_T(t)$ is the hazard function (or risk function or mortality rate). In the classical theory, the initial value for reliability, $L_T(0)$, is usually assumed to be 1.0. Suppose that the revised reliability following ith inspection is given as follows:

$$L_T^{(i)}(t) = L_T(t_i) \exp\left[-\int_{t_i}^t h_T^{(i)}(t)\mathrm{d}t\right] \tag{12}$$

where $L_T(t_i)$ can be considered as the up-dated estimate for the safety state of the particular structure using inspection results and test data. It can be related to the safety measure as given in Eq. 10 as follows:

$$L_T(t_i) = 1 - 10^{-N_i} \tag{13}$$

where the subscript denotes the corresponding inspection from which this measure is estimated. Currently available methods of inspection and testing are briefly reviewed in Chapter 3.

The second problem is the estimation of the hazard function $h_T(t)$ as given in Eq. 11, and its modification after we obtain inspection and test results. Several methods currently under investigation are also summarized and discussed in Chapter 3.

3 Safety evaluation of existing structures

3.1 General remarks

In Chapter 2, several definitions of structural damage were reviewed. In this chapter, several currently available approaches in structural engineering practice are summarized and discussed. In addition, some methods as proposed by other investigators are also given.

3.2 Current approaches in engineering practice

In cases where statistically sufficient data are available, the problem of safety evaluation can be formulated with the application of the theory of probability and statistics. For example, Rudd *et al.* (108) presented a statistically-based methodology for the quantification of structural damage due to cracking in mechanically fastened joints. They successfully presented an analysis for a lap shear specimen subjected to a B-1 bomber load spectrum. As another example, Berens *et al.* (10) presented results of a comprehensive statistical study of crack growth. The unique feature of most civil engineering structures is that they are not only much larger and more complex than aircraft structures but also design dependent. In the great majority of cases, the same design is used and built only once. Therefore, it is not economically feasible to collect statistically sufficient data. In fact, the structural behavior near failure remains unknown for most civil engineering structures because it is not practical to conduct even a single full-scale destructive test of such large and massive structures.

In civil engineering practice, a given structure can be studied either experimentally or analytically or both whenever one or more interested so wish as (*a*) a result of observing signs of distress or failure or (*b*) a part of periodic inspection procedure (65). The experimental studies can be field surveys and/or laboratory tests. Field surveys include the determination of exact locations of failed elements and other signs of distress, the conduct of various nondestructive testing techniques to the existing structure, the discovery of poor workmanship and defective details, and proof-load and other load testing of a portion of a large structure. On the other hand, specimens can be collected from the field and tested in the laboratory for the evaluation of material strengths and other mechanical properties. Analytical studies may consist of the examination of the original design calculations and drawings, the review of project specifications, the performance of additional analyses using newly acquired field observations and test data, and the possible explanation and description of the event under consideration.

In 1981, Guedelhoefer (64) presented a comprehensive summary of methods which are applicable for the strength evaluation of distressed structures. He classified these methods into the following three types: (*a*) analytical evaluation, (*b*) in-situ load testing, and (*c*) model load testing. Moreover, he suggested that it is necessary to follow an integrated systematical approach involving the following three frequently iterating steps: (i) condition survey, (ii) structural evaluation, and (iii) correlation. The condition survey should produce a comprehensive documentation of the in-situ and as-built condition of the structure being evaluated. The structural evaluation is the determination of the structural behavior including the effects of all relevant field conditions. The correlation step is the description of the relationship between predicted structural behavior with the observed distress conditions. For detailed descriptions of various methods being used, readers should read the original paper by Guedelhoefer (64).

I am privileged to have read several proprietary reports concerning the investigation of distressed structures. Even though I understand the general procedures and various methods as applied by particular investigators, the decision-making process remains as privileged information for a relatively few experts and are transmitted to younger engineers primarily through experience and intuition. In the following section, several methods resulting from such experience and intuition are summarized.

3.3 Several proposed methods

The methods of Wiggins and Moran (133), Culver *et al.* (36), and Bresler and Hanson (17) as reviewed in Section 2.2 may be included here as two of these proposed methods. In the mid-seventies, a safety evaluation program was developed (18, 85). The structural and fire evaluation model (SAFEM) was developed to provide a broad overview of potential safety problems for more than 10,000 buildings for a governmental agency in the States. A building can be classified into (*a*) 'green' requiring only routine scrutiny, (*b*) 'yellow' requiring some attention, and (*c*) 'red' requiring immediate attention and improvement. Authors emphatically stated that

'SAFEM is not a substitute for an engineering analysis, but it directs attention to buildings which require engineering analysis on a priority basis'. The procedure consists of (i) collection of such data as building size, cost, number of occupants, address, and predetermined exposure to natural hazards; (ii) ranking buildings on the basis of priorities; (iii) choosing buildings which should undergo field surveys; (iv) performing field surveys and recording survey results in the computer file; (v) re-ranking buildings on the basis of priorities and requesting engineering studies for buildings with the largest potential problems, (vi) performing engineering studies and producing the final priority rankings, and (vii) allocating funds for upgrading these structures following these priorities. A detailed computer program is developed on the basis of professional experience to combine numbers ranging from 0 to 9 for hazards (geophysical, intrinsic, and local), exposure, and vulnerability. The SAFEM profiles include one each on fire, structural, and miscellaneous (glass safety, cladding failure, electrical system, elevator system, etc.). Although this program covers a broad scope and many detailed considerations, it is difficult to understand how the computer program evolved with the many subjective inputs.

Several recent damage assessment studies are reviewed in a comprehensive manner by Scholl *et al.* (111), who recommended a general procedure. By computing the maximum floor responses, they estimate the component damage by using a component motion-damage library (86). The damage to a structure is defined as the sum of the component damage which in turn are sums of sub-component damage.

Recently, Hart *et al.* (67) proposed the use of reliability indices for the evaluation of structural damage. Their evaluation process consists of the following parts: (*a*) identification of possible failure modes, (*b*) selection of analytical models representing these failure modes, (*c*) quantification of the mean values and uncertainties for respective loads and resistances, (*d*) calculation of the reliability index for each failure mode, and (*e*) application of decision theory techniques to evaluate the potential structural damage. FitzSimons (50) also suggested the use of two techniques which are related to statistics and uncertainties.

Meyer *et al.* (98) reported on their development of an analytical procedure for the assessment of structural safety. Specifically, their investigation consisted of the following tasks: (*a*) establishment of a stochastic model representing the seismic environment of the building site; (*b*) establishment of a mathematical model simulating the structural response to strong earthquake excitations; (*c*) definition of damage parameters suitable for reliability analyses, and (*d*) reliability analysis of the building for its safety evaluation. They used a modified damage probability matrix approach as originally proposed by Whitman *et al.* (134, 135). With the assumption that a correlation can be established between the MMI (Modified Mercalli Intensity) value, the severity of concrete cracking and the degradation of structural stiffness, Meyer *et al.* (98) estimated the expected damage for any given MMI value or determined the MMI value from a visual inspection of the structure. To illustrate this method, they considered a concrete beam and used a Markovian transition matrix and Monte Carlo simulation techniques to obtain numerical results as presented in their paper (98).

Using information as found in historical documents on the Boston Old State House, Luft and Whitman (91) estimated an upper bound for the peak ground acceleration of the 1755 Cape Ann earthquake using structural analysis techniques. They collected and analyzed such historical records as earthquake damage to the building, structural configuration, methods and materials of the construction, and local soil conditions. Then, they developed a structural model, with which a structural analysis was made. The levels of base acceleration was predicted and correlated with the recorded damage.

Sues *et al.* (122) presented a method for the safety evaluation of structures subjected to earthquake hazards. They measure the safety level in terms of the probability of exceeding a specified level of damage, which may be defined in terms of any response variable. A seismic hazard model is used to calculate the probabilities of all relevant ground motions over a specified time period. Moreover a hysteretic structural model is used to calculate the required statistics of its random response to these expected earthquakes. The uncertainties in structural modeling and ground motions are considered in the calculation of conditional probabilities of damage for given earthquake intensities. These conditional probabilities are then combined with appropriate seismic hazard probabilities to obtain the desired probability of structural damage. For the purpose of illustration, a hypothetical four-story building was modeled as a four-degree-of-freedom shear-beam system. They found that the calculated damage probabilities represent a rational assessment of the seismic safety of a structural design.

3.4 Discussion

An attempt is made to review current approaches being used in the practice. Because (*a*) proprietary information is involved and (*b*) the real-world problems are still too complex to be understood completely, only general investigation procedures and obvious case histories are described in the open literature (68, 87).

Several proposed methodologies are also reviewed herein. Each one of these methods is logical and rational. However, it is necessary to provide reasonable input values or to interpret available information in order to obtain meaningful results from using these methods. It seems that more work is needed to make these methods useful in a practical sense. It is also desirable to compare results of applying different methods to a few standard structures such that these methods can be calibrated with one another. Eventually, it is desirable to develop a series of safety evaluation procedures for various types of structures under different circumstances. These and other methods can be included as parts of such procedures.

4 System identification in structural dynamics

4.1 General remarks

It is well known that the mathematical models as used in structural analysis and design prior to the construction phase do not truly represent the behavior of a given structure. To obtain improved mathematical models for a better simulation of the real structure, response records with or without known forcing functions have been collected and analyzed with system identification techniques during these past two decades. By necessity, these tests are usually conducted at small response amplitudes so that the serviceability and safety limit states are not exceeded. Consequently, the resulting modified mathematical models are applicable to the linear or slightly nonlinear range of the structural behavior. At present, it is possible to simulate the structural response to extreme forces such as strong earthquakes or wind storms with the use of digital or hybrid computers, and thus to evaluate the serviceability and safety conditions of the structures. Consequently, there exists the paradox that (*a*) the applicability of 'realistic' models of the structure are limited to small-amplitude response range, (*b*) the catastrophic loading conditions are likely to cause the structures to behave other than the linear or 'near-linear' responses which are usually assumed, and (*c*) the severe loadings may cause serious damages in the structure and thus change the structural behaviors appreciably. It is important that the extent of damage in structures can be assessed following each major catastrophic event or at regular intervals for the evaluation of aging and decaying effects. On the basis of such damage assessment, appropriate decisions can be made as to whether a structure can and should be repaired to salvage its residual values.

The objectives of this chapter are to (*a*) review and summarize the available literature on the methods of system identification in structural dynamics and (*b*) discuss how the results of system identification studies may be used to obtain a rational procedure for the safety evaluation of existing structures.

System identification is a process for constructing a mathematical description or model of a physical system when both the input to the system and the corresponding output are known. For most of the current applications, the input is usually a forcing function and the output is the displacement or other motions of the structure subjected to these forces. The mathematical model obtained from the identification process should produce a response that in some sense matches closely the system's output, when it is subjected to the same input. In general, the system identification technique consists of the following three parts: (*a*) determination of the form of the model and the system parameters; (*b*) selection of a criterion function by means of the 'goodness of fit' of the model response to the actual response that can be evaluated, when both the model and the actual system are subjected to the same input; (*c*) selection of an algorithm for modification of the system parameters, so that the discrepancies between the model and the actual system can be minimized. The techniques for modeling and numerical calculations have been developed to a high degree of sophistication in all branches of engineering. Particularly, in the areas of electrical and mechanical control system analysis, the identification techniques have found wide ranges of practical application. However, these techniques cannot be readily applied to structural analysis. Because of the large size and mass of most real structures, many common techniques for generating a convenient force input, and hence a suitable system output, are no longer practical for the identification of civil engineering structural systems. Only limited source of input, such as vibrations due to earthquakes, strong wind loads, controlled explosions, are possible to generate sufficiently large excitation. Even for laboratory simulations, the limitations on the types of structure and the types of response which can be performed in a laboratory are far greater than an electrical system or a mechanical system. In addition, most of the inputs and outputs are random in nature. To extract useful information from these data poses a new problem to system identification studies in other disciplines.

4.2 System identification techniques

System identification techniques have been widely used in many branches of science and engineering to identify various characteristics of a physical system (47, 109). Their applications in civil engineering structures have drawn much attention only in the last two decades. The primary objective of applying system identification in structural engineering is to obtain a mathematical model which can best represent the characteristics of the structure. In the available literature, a set of differential equations (lumped-mass model or simple continuous model) or a transfer function (black box model or lumped-mass model in frequency doman) is often used to represent the structural behavior. A set of parameters is to be estimated from the measured response of the real structure to a known disturbance. The application of system identification techniques to solve structural engineering problems is called structural identification by several investigators (25–28, 69, 96, 97, 105, 106, 121, 125, 126).

Because of their simplicity, the linear lumped-parameter models are the most widely used models in structural identification. More complex models such as the linear continuous-parameter models and non-linear-parameter models are used only when the lumped-parameter model cannot be used to provide an adequate representation of the structural behavior. For lumped systems or continuous systems with lumping approximations, the disturbance must also be represented in a discrete form. On the other hand, the disturbances can be either discrete or continuously distributed in a continuous system.

Note that parameters in a lumped system need not have physical meanings. In a mathematical representation, the geometry, material properties, interactions between various structural elements, boundary conditions, etc., are all 'lumped' into the parameters assumed. Thus, the parameters are combined empirical indices, which are valid only for the particular excitation and structural response as used in the identification process. To extend the model to include some physical inputs, continuous models have been assumed to give a more rational approximation of the real structural behavior. These models are usually formulated in the form of differential or integral equations. For numerical calculations, the equation is usually discretized by using the finite-difference techniques. Then the system is again reduced to a discrete-parameter system. In the subsequent numerical calculation, such a system is usually more difficult to analyze as compared to that involved in a direct lumped-mass model. An alternative form of discretization involves the use of the finite-element method to represent a real structure. The versatility of the finite-element method may prove to be most advantageous when two- or three-dimensional structural problems are considered. Its application in the structural identification has only been explored very recently.

In the method of modal expansion, the structural responses such as displacements are expressed in terms of the shape functions for the normal modes. The equations of motion describing the structural model are usually decoupled, and the formulation can be written in terms of generalized coordinates. The solutions (i.e. parameter values) are readily available (30, 105, 106). It is also possible to extend the method to problems involving non-proportional damping with expansions in terms of non-normal modal shape functions (78).

It is convenient to define the physical characteristics of a structural system in the frequency domain. A transfer function, defined as the ratio of the response function to the excitation function in the Laplace or Fourier frequency domain, is usually taken to represent the structural model for linear and time-invariant system. The physical interpretation of the inversion of a transfer function may be taken as the response of a structure to a unit impulse. The transfer function is usually rewritten in an algebraic form with coefficients to represent the combined effects of spring constants, masses, viscous coefficients of a linear spring-mass structural behavior model. Because the functional form and the coefficients have no direct physical correspondence, it is generally called a 'black box' approach. Estimation based on the finite Fourier transformation has the advantage of minimizing truncation errors. Fast Fourier transformation provides an appreciable reduction of computation time and reduces the round-off errors (69, 77, 78).

Various least-squares estimation methods (including repeated and generalized least squares), the instrumental variables method, the maximum likelihood estimation, and the tally principle have been used to identify linear structural models. The least-squares estimation minimizes the summation of square errors between the predicted response and the measured structural response. In the generalized least-squares method, the criterion function for evaluating the 'goodness-of-fit' is the summation of square generalized errors which is defined to include the additive noise covariance matrix. Repeated application of the least squares method can be used to modify the usual least squares procedure by increasing the order of the mathematical model in an iterative process until the desired accuracy is achieved. Although the validity of these methods has not been proved formally, they have been applied to structural identification problems with satisfactory results.

The instrumental variables method is applicable to the problem of bias with noise-polluted responses. The method involves an iterative process in the calculation of revised estimate and instrumental variables matrix function. The maximum likelihood method is widely used for parameter estimation in statistics. It determines the parameter estimate by minimizing the criterion function through an iterative procedure. The method appears to have the advantage of providing the best estimation for a wide range of contamination intensity in the external excitation and the structural response (63). The estimation methods are generally applied to the time-domain analyses, and usually involve complicated iterative procedures. Nevertheless, they can be used to treat nonlinear models for which the modal expansion and transfer functions in frequency domain are not defined.

In contrast to the work done on linear models, relatively little seems to have been done on nonlinear models. It is in part due to the mathematical difficulties in considering the nonlinear terms. Some common techniques in dealing with linear systems, such as the modal expansion and transfer function, are not appropriate in the nonlinear case, though it is possible to apply the modal expansion analysis to obtain approximate solutions for slightly nonlinear problems. It is also because the current developments in structural identification have mostly dealt with structural parameters with limited range of application or parameters for highly simplified structural behaviors. For example, in the evaluation of vibratory parameters of structures, the models are often limited to small-amplitude response range and time-invariant structural behaviors. However, the catastrophic loading conditions such as strong earthquakes and windstorms are likely to cause the structure to behave beyond the linear range of responses which are usually assumed. More importantly, the severe loadings may cause serious damage in the structure and thus change the structural behaviors appreciably.

Using the theory of invariant imbedding, a best *a priori* estimate can be obtained by minimizing an error function (39). The method is applicable to some general boundary conditions. Dynamic programming filter is a more general method with the invariant imbedding as a special case. Instead of going through the Euler–Lagrange equations to determine the best estimate that minimizes the error function, dynamic programming may be applied directly. In such cases, the decomposition of the error function can lead to a system of partial differential equations. The optimal least squares filter satisfies the governing differential equation which describes the structural model and minimizes the quadratic error function. The error function is defined in terms of observed error vectors (weighting matrices) and the best *a priori* estimate of the parameters (39–44).

The Gauss–Newton method belongs to the general family of quasilinear methods based on a linear expansion of the system variable around an available estimate of the variable. However, the convergence is not guaranteed (42). If the calculation is convergent, it converges quadratically.

If accurate acceleration measurements are available, a direct approach may be used without an initial estimate of the coefficients. In some cases, it requires only partial estimates. The parameters are determined by directly minimizing the quadratic error function. The method appears to be efficient in computation, particularly for nonlinear models with a single degree-of-freedom (42).

The Kalman filter has been used to obtain optimum sequential linear estimation and an extended filter deals with nonlinear filtering. Its good approximation for high sampling rates has been demonstrated in simulation studies of parameter estimation (109).

The maximum likelihood method has been applied to both linear and nonlinear systems. It can be used to treat both the measurement noise and the process noise, and may also be used to estimate the covariances of the noises (105). It is also suggested that the extended Kalman filter may be introduced in the calculation of the likelihood function.

An input–output relationship of multiple integral form is assumed to represent the model (92). The kernel functions which represent model parameters can be estimated by a cross-correlation technique. In theory, the relationship can be written in Laplace domain and thus the kernels are identified in terms of the Laplace parameter. Their values in the time domain are then obtained by the usual inversion techniques.

Most of the literature which has been surveyed in this chapter deals with the linear lumped-parameter model or the linear continuous model. The formulation is given in the form of a set of linear equations of motion:

$$m\ddot{x} + c\dot{x} + kx = F \tag{14}$$

where x is the structural displacement response matrix, F is the excitation matrix (usually the external forces), m is the mass matrix, c is the damping matrix and k is the stiffness matrix. Hence, the parameters to be identified are usually the m, c and k matrices. As discussed previously, these matrices are not necessarily related to the mass distribution and material stiffness of the real structure.

The form of nonlinear models generally varies with the type of excitation and the algorithm employed for numerical calculation. One of the direct extensions of the linear model can be obtained by assuming that

$$m\ddot{x} + g(\dot{x}, x) = F \tag{15}$$

where the nonlinear function g may be taken as an odd algebraic function of $\dot{x}$ and x (42, 93, 94, 125–128), e.g.,

$$g(\dot{x}, x) = a_1\dot{x} + a_2\dot{x}^3 + a_3x + a_4x^3 \tag{16}$$

The integral form of the formulation of the excitation–response relationship has also been used when transfer function is used for the linear model. In an integral formulation, instead of using three constant-parameter matrices, i.e., m, c and k, the model characteristics are lumped in a kernel function $h(\tau)$ in the following form:

$$x(t) = \int_0^\infty h(\tau)F(t-\tau)\mathrm{d}\tau \tag{17}$$

It is easy to extend the integral formulation to include the nonlinear kernels. For example, a second-order model has the form (92, 131),

$$x(t) = \int_0^\infty h_1(\tau)F(T-\tau)\mathrm{d}\tau + \int_0^\infty\int_0^\infty h_2(\tau_1, \tau_2)x(t-\tau_1)x(t-\tau_2)\mathrm{d}\tau_1\mathrm{d}\tau_2 \tag{18}$$

However, the computation efforts involved for the second or higher order models are much greater. Chen *et al.* (25, 125) classified the types of forcing function, structural model, and structural response in their survey of some 94 references. Collins, Young and Kiefling (33) surveyed the system identification techniques in the shock and vibration area. A technology tree was developed along two principal branches—the frequency domain and the time domain—for the purpose of assisting engineers in matching a particular need with available technology. Specific examples of accomplished activity for each identification category were discussed. Special emphasis was focused on the use of statistical approach in structural identification (31, 32). Numerical examples for the estimation of the stiffness of a spring-mass chain and a two-degree-of-freedom system were given by using the weighted least squares method.

Hart and Yao (69) presented a review of the identification theories and applications in structural dynamics as of 1976. They included identification problems which require a prior structural model with or without a quantification of experimental and modeling errors. The review also contained a brief description of the algorithms and sample data.

Later, Liu and Yao (90) discussed the concept of structural identification in the context of system identification and unique characteristics in its structural engineering applications. Basically, structural engineers are interested in identifying the damage and reliability functions, respectively $D(t)$ and $L(t)$, in addition to the equation of motion. From another viewpoint, the updated equation of motion using test data and system identification can be a tool for the estimation of damage and reliability of existing structures as shown in Fig. 3.

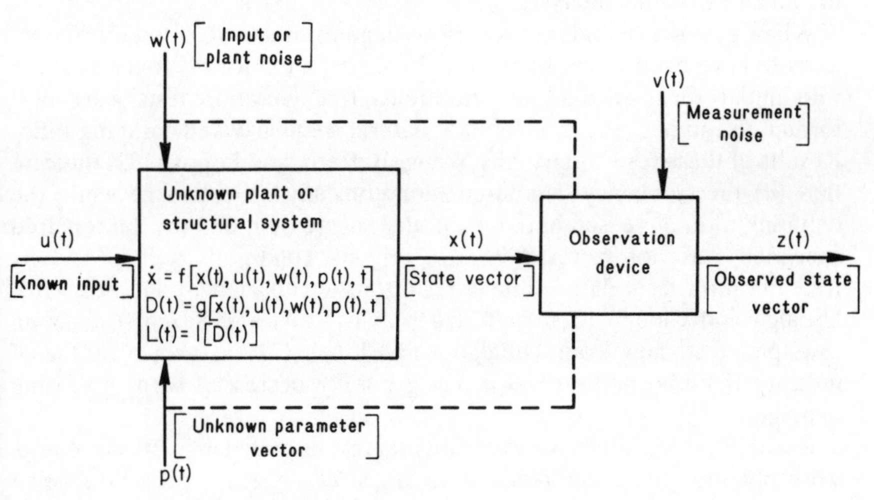

Fig. 3 General structural identification problem (90)

When a structure is inspected for the purpose of making damage assessment, a sequence of tests may be conducted and the resulting data can be analyzed accordingly. Quantities which can be measured and recorded in testing structures include the load, the deformation (or strain) and the acceleration. From such data, mechanical properties such as stiffness and strength and dynamic characteristics such as natural frequency and damping can be estimated. In addition, visible damage such as cracks and local buckling in the plastic range can be detected by experienced observers. As a practical example, binoculars have been used by persons looking for color change in window panes in a tall building, which indicates the presence of flaws causing the eventual breakage of window glasses.

For metal structures which are subjected to repeated load applications, dye-check, ultrasonic or X-ray devices may be used to find and measure small and hidden fatigue cracks which indicate structural damage. The effect of detecting such fatigue cracks during a periodic inspection on the structural reliability of aircraft structures was studied by Yang and Trapp (138).

Many full-scale on-site load tests of building structures have been performed in this country during the past several decades (51). To date, most field load-tests are static in nature and limited to studies of flexural response. FitzSimons and Longinow (51) emphasized the fact that a static test cannot be used to reveal such weaknesses of a given structure as those due to corrosion, repeated load, creep, and brittleness. Nevertheless, load tests can be used to improve the reliability estimate (116). Moreover, valuable information such as the stiffness of the structure can be obtained for the improvement of the mathematical representation of the structure for further dynamic analysis.

When a structure undergoes various degrees of damage, certain characteristics have been found to change. In testing a reinforced concrete shear wall under reversed loading conditions, free vibration tests were performed to estimate the fundamental natural frequency and damping ratio. Results of these tests as given by Wang, Bertero, and Popov (132) indicate that (*a*) the frequency decreased monotonically with damage while the damping ratio increased initially and then decreased, and (*b*) the repaired specimen was not restored to the original condition as indicated by free-vibration tests data. Similar results were reported by Hudson (76), Hidalgo and Clough (72), and Aristizabal-Ochoa and Sozen (6). As an example, test data from Hidalgo and Clough (72) as shown in Fig. 4 indicate that the natural frequency generally decreased with increasing damage.

Recently, comprehensive experimental results of dynamic full-scale tests were obtained for a multi-story building structure (61) and a three-span highway bridge (7). Galambos and Mayes (61) tested a rectangular eleven-story reinforced concrete tower structure, which was designed in 1953, built in 1958, and tested in 1976. The large-amplitude (and damaging) motions were induced with the sinusoidal horizontal movements of a 60-kip lead-mass which was placed on hardened steel balls on the eleventh floor. This lead-mass can be displaced up to ±20 inches and the frequency

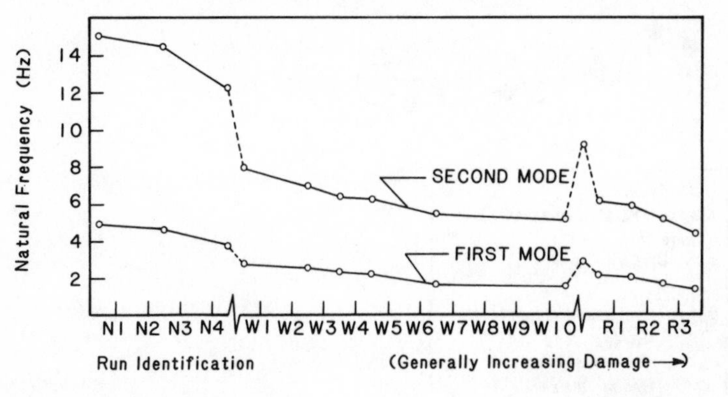

Fig. 4 Natural frequency variation through test history (after Hidalgo and Clough (72))

capacity was 5 Hz with the use of a servo-controlled hydraulic actuator, one end of which is fastened to the building frame. The maximum horizontal force range was ±30,000 pounds. Some test results are plotted as shown in Fig. 5 indicating that the natural frequency decreased with increasing damage in general. Similarly, Baldwin *et al.* (7) concluded from their testing of a three-span continuous composite bridge that changes in the bridge stiffness and vibration signatures can be used as indicators of structural damage under repeated loads. Further analyses of such full-scale test data should be very useful in understanding the structural behavior as well as in making damage assessment of existing structures.

System identification tests are always conducted at extremely low-level vibrations; they can be performed as many times as it is needed without causing any apparent damage to the structure. In most cases, only earthquake records are available for the purpose of damage analysis. In general, the complete record of an earthquake can be separated into the following three portions with different characteristics: (*a*) strongly-excited portion with higher modes contribution at the beginning of an earthquake, (*b*) much larger amplitude portion with nonlinear behavior, and (*c*) very low level vibration portion at the end of an earthquake. In system identification problems, parameters identified from portion (*a*) cannot be very accurate although a contribution of higher modes has been considered in the analysis because of higher irregularity of earthquake input and response data. Because portion (c) is equivalent to very low level ambient vibration, natural frequency identified from portion (c) is always higher than and cannot be compared with that from portion (b). However, the period of portion (a) and the relatively low amplitude of portion (c) cannot be determined because they depend on the duration and intensity of earth-

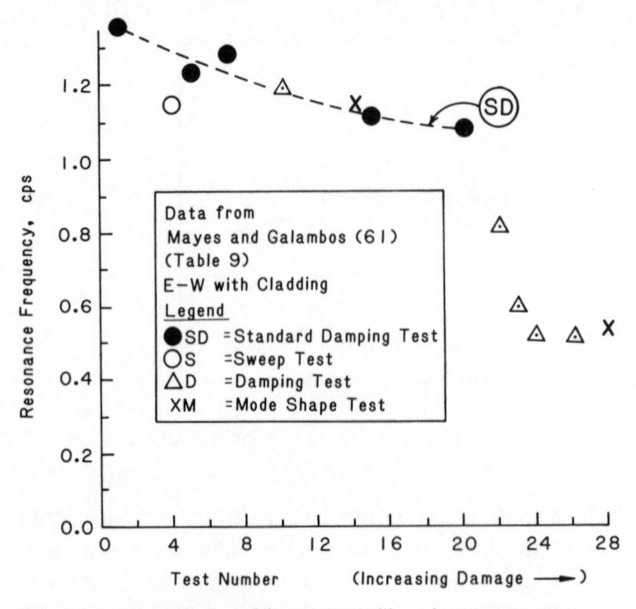

Fig. 5 Variation of natural frequency with a damage measure

quake and structural characteristics. Consequently, one approach used by Chen (26–28) is to deal with the identification of structural characteristics only in portion (b) by dividing this portion into several segments in order to study and compare the changes among those characteristics. Method I is applied to find parameters ω_n (natural frequency) and ξ (damping coefficient) as functions of time from two linear equations of motion at time t and time $t + \Delta t$ by using measured earthquake and response data, $x_0(t)$ and $y(t)$, respectively. Parameters at any time t can be found as follows:

$$\omega_n^2(t) = \frac{(\ddot{x}_0(t) + \ddot{y}(t))\dot{y}(t + \Delta t) - (\ddot{x}_0(t + \Delta t) + \ddot{y}(t + \Delta t))\dot{y}(t)}{\dot{y}(t)y(t + \Delta t) - \dot{y}(t + \Delta t)y(t)} \tag{19}$$

and

$$\xi(t) = \frac{\ddot{x}_0(t) + \ddot{y}(t) + \omega_n^2(t)y(t)}{2\omega_n(t)\dot{y}(t)} \tag{20}$$

When the denominators of Eqs. 19 and 20 approach zero due to inadequate choice of Δt, nonlinear behavior, higher mode contribution, and measurement noise, it is difficult to estimate the values of $\omega_n(t)$ and $\xi(t)$ using this method. By using the least-squares-error-fit, natural frequency, ω_n, and damping ratio, ξ, can be estimated by minimizing the following integral-squared difference, E, between the excitation, x_{oi}, input to a structure and the excitation, $\ddot{x}_{oi}^{(1)}$, calculated from its linear model:

$$\begin{aligned} E &= \sum_{i=n_j}^{n_k} (\ddot{x}_{oi} - \ddot{x}_{oi}^{(1)})^2 \\ &= \sum_{i=n_j}^{n_k} (\ddot{x}_{oi} + \ddot{y}_i + 2\xi\omega_n\dot{y}_i + \omega_n^2 y_i)^2 \end{aligned} \tag{21}$$

Chen (26) obtained the following estimates:

$$\omega_n^2 = \frac{(\Sigma\ddot{y}_i\dot{y}_i + \Sigma\ddot{x}_{oi}\dot{y}_i)\Sigma\dot{y}_i y_i - (\Sigma\ddot{y}_i y_i + \Sigma\ddot{x}_{oi} y_i)\Sigma\dot{y}_i^2}{\Sigma\dot{y}_i^2\Sigma y_i^2 - (\Sigma\dot{y}_i y_i)^2} \tag{22}$$

and

$$\xi = -\frac{\Sigma\ddot{y}_i y_i + \Sigma\ddot{x}_{oi} y_i + \omega_n^2\Sigma y_i^2}{2\omega_n\Sigma\dot{y}_i y_i} \tag{23}$$

where $\ddot{x}_{oi}$, $\ddot{y}_i$, $\dot{y}_i$, and y_i are measured data during an earthquake, and intervals n_j, n_k are segments of the records to be used for identification of ω_n and ξ. In addition, values ω_{n1} and ξ_1, ω_{n2}, and ξ_2, . . ., ω_{nn} and ξ_n are identified from segments (n_1, n_2), (n_2, n_3), . . ., (n_n, n_{n+1}), respectively. Note that Eq. 22 and Eq. 23 are reduced to Eq. 19 and Eq. 20 when only two points of record are taken in least-squares-error-fit.

The main advantage of using such simple methods and linear models is the ease in checking the results and in preventing unnecessary errors associated with complicated calculation. In a possible future application to structural control (142), such simple methods can be very useful. Moreover, the exact nonlinear characteristics for any given full-scale structure are not readily known in most cases.

These methods are applied to analyze response records of two buildings collected during the 1971 San Fernando Valley earthquake (52). The Union Bank Building is a 42-story steel-frame structure in downtown Los Angeles. Prior to the 1971 San Fernando earthquake, strong-motion accelerographs with synchronized timing were installed in the sub-basement, on the 19th floor and on the 39th floor. However, the instruments on the 39th floor failed to function. The S38°W components of the digitized relative acceleration, velocity and displacement at the 19th floor were used as the response data in the analysis. As shown in Fig. 6, the results of Chen's methods agreed with those of modal minimization method by Beck (8, 9). The natural frequency is shown to decrease segment by segment except for the last value obtained from Method II. The loss of stiffness as indicated by this change in natural frequency seems to be the result of cracking and other types of damage in nonstructural elements during the occurrence of large-amplitude earthquake response.

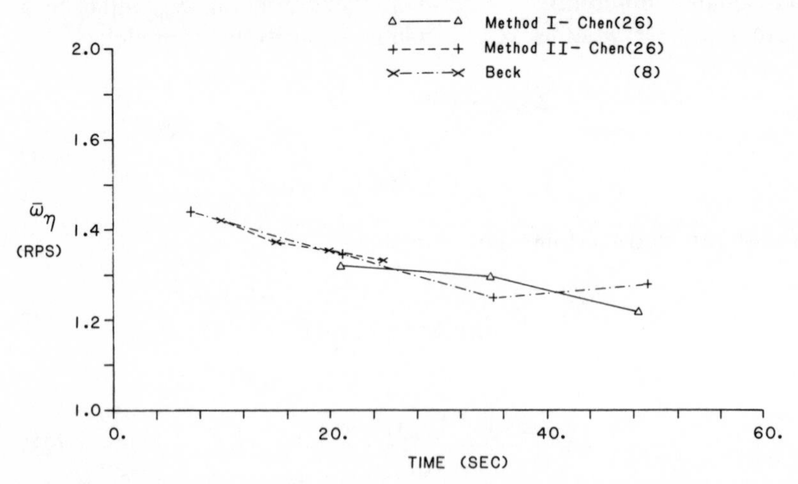

Fig. 6 Comparison of the natural frequency identified from different methods for Union Bank building

Building 180 is a nine-story steel-frame structure on the grounds of the Jet Propulsion Laboratory, Pasadena, California. The S82°E components of the ground acceleration, relative acceleration, velocity and displacement at the roof were used as the excitation and response data. Figure 7 shows

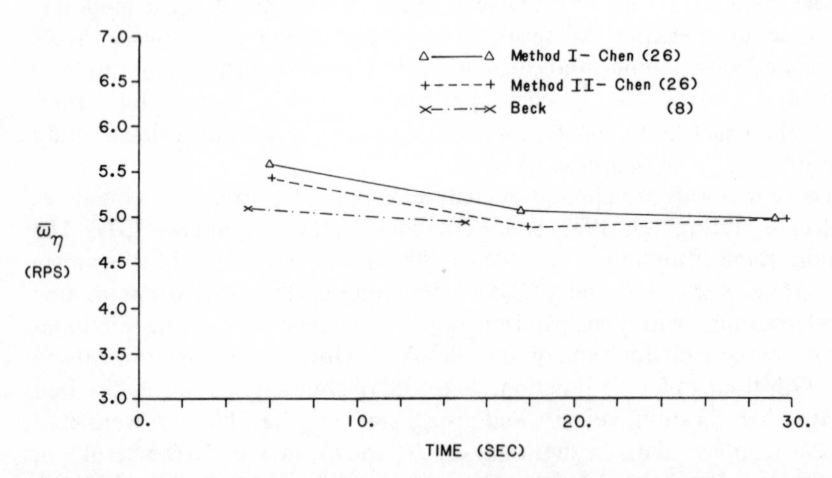

Fig. 7 Comparison of the natural frequency identified from different methods for JPL building 180

results from Method I and Method II by Chen (26), and Modal Minimization method by Beck (8, 9). The amplitude of the acceleration response of this building during the earthquake was twice that of the Union Bank but damage was limited to minor nonstructural cracking. However, very little changes in the natural frequency result in this analysis due to relatively minor damage involved.

4.3 Application of non-parametric methods to structural dynamics

Consider a single-degree-of-freedom system as shown in Fig. 8(a). Its spring and damping forces are assumed to be functions of displacement and velocity response, respectively, as shown in Fig. 8(b). Suppose that the displacement and velocity response can be obtained from recorded data as shown in Fig. 8(c). The local maximum and minimum values of the

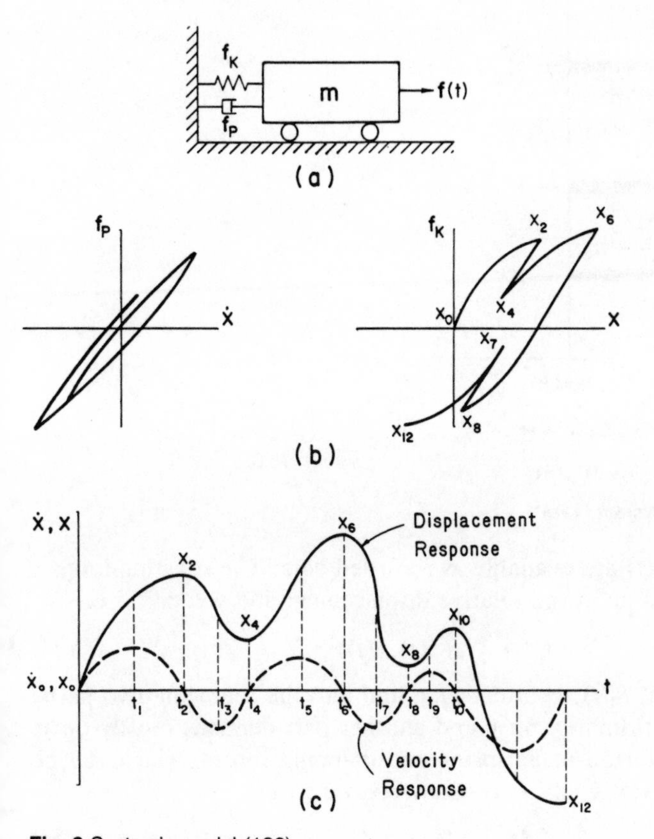

Fig. 8 System's model (126)

displacement occur whenever the velocity response is zero. In between these local maximum and minimum values, polynomial functions such as Eq. 16 can be used to describe the relationships between (*a*) the damping force f_P and velocity $\dot{x}$ and (*b*) the spring force f_K and displacement x (39, 42, 93, 94, 126–128, 130).

Toussi (126) developed such a nonparametric method for a multi-story building frame as shown in Fig. 9, where the relative acceleration $y(t)$ and

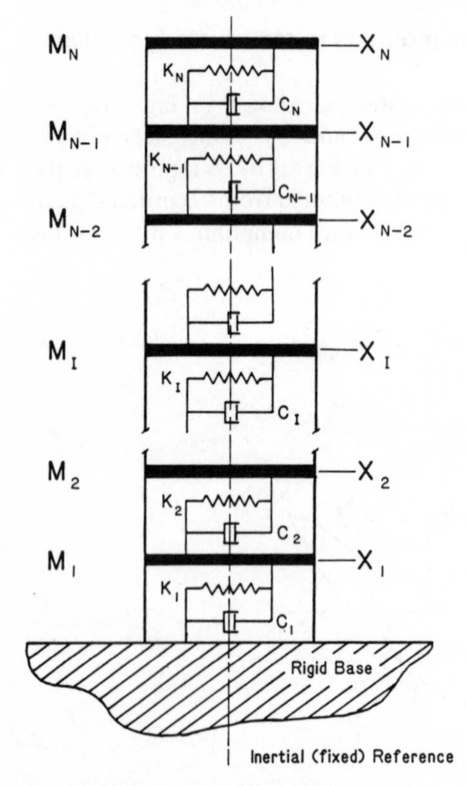

Fig. 9 Lumped-mass system (126)

the applied force $f(t)$ are available as recorded data. The resisting force is assumed to depend upon the relative displacement and velocity, i.e.,

$$f_R(t) = f_R(\dot{y}(t), y(t)) \tag{24}$$

The resisting force, $f_R(t)$, is often separated into the following two parts: one part depends primarily on $\dot{y}$ and another part depends mostly on y. Here, they are referred to as spring and damping forces, which can be nonlinear functions of $\dot{y}$ and y, respectively, i.e.,

$$f_R(t) = f_P(\dot{y}) + f_K(y) \tag{25}$$

where the damping force, $f_P(\dot{y})$, and the spring force, $f_K(y)$, are defined as follows:

$$f_K(y) = \sum_{j=0}^{p} a_j y^j(t) \tag{26}$$

$$f_P(\dot{y}) = \sum_{j=0}^{q} b_j \dot{y}^j(t) \tag{27}$$

and

$$p + q \leqslant n_i - 1 \tag{28}$$

in which n_i indicates the number of points (samples) within interval i. Substitution of Eqs. 25 and 26 into Eq. 24 yields

$$f_R(\dot{y}, y, t) = c + \sum_{j=1}^{p} a_j y^j(t) + \sum_{j=1}^{q} b_j \dot{y}^j(t) \tag{29}$$

where

$$c = a_0 + b_0 \tag{30}$$

Repeating Eq. 28 for the response components of each one of these n_i points results in n_i simultaneous equations as follows:

$$F_R = YA \tag{31}$$

where F_R, Y and A represent the resisting force vector, response matrix and parameter vector of the ith interval respectively. The elements of vector F_R are calculated by setting the equilibrium of forces which are acting on the structural system equal to zero.

$$f_R(t) = my(t) + f(t) \tag{32}$$

Because y and $\dot{y}$ can be calculated by integrating the recorded acceleration, $\ddot{y}$, the response matrix, Y, is also known. Therefore, the vector A can be found by pre-multiplying the inverse of matrix Y on both sides of Eq. 30. Thus

$$A = Y^{-1} F_R \tag{33}$$

after the parameter vector is calculated, the components of the resisting force can be obtained.

It is important to study the measuring and recording noise and the effect of imperfect mathematical representation of structural behavior. The presence of noise suggests the use of mathematical statistics. Because statistical calculations require a sufficient number of samples, instead of increasing the number of samples, the number of parameters is reduced in Toussi's study (126).

Figure 9 shows the lumped-mass model for a building with masses m_j, $j = 1, 2, \ldots, N$. It is assumed that these masses are connected by nonlinear dashpots C_j, and nonlinear springs, K_j, as shown. The recorded motions consist of the acceleration of the base, $\ddot{y}_g(t)$, and absolute acceleration of the floors, $\ddot{x}_j(t), j = 1, 2, 3, \ldots, N$. The absolute velocity and displacement of each floor is obtained by integrating the corresponding absolute acceleration. The forces created in the springs and dashpots are assumed to depend upon the relative displacement and velocity between the neighboring masses, respectively, i.e.,

$$[f_K(t)]_j = f_K[Y_j(t)], \quad j = 1, 2, \ldots, N \tag{34}$$

$$[f_P(t)]_j = f_P[\dot{Y}_j(t)], \quad j = 1, 2, \ldots, N \tag{35}$$

where

$$Y_j(t) = x_j(t) - x_{j-1}(t), \quad j = 1, 2, \ldots, N \tag{36}$$

$$\dot{Y}_j(t) = \dot{x}_j(t) - \dot{x}_{j-1}(t), \quad j = 1, 2, \ldots, N \tag{37}$$

Now the equilibrium of forces applied to each floor is formed as follows:

$$\begin{aligned}
&m_N\ddot{x}_N(t) + f_P[\dot{Y}_N(t)] + f_K[Y_N(t)] = 0 \\
&m_{N-1}\ddot{x}_{N-1}(t) + f_P[\dot{Y}_{N-1}(t)] + f_K[Y_{N-1}(t)] = f_P[\dot{Y}_N(t)] + f_K[Y_N(t)] \\
&\cdot \\
&\cdot \\
&\cdot \\
&m_j\ddot{x}_j(t) + f_P[\dot{Y}_j(t)] + f_K[Y_j(t)] = f_P[\dot{Y}_{j+1}(t)] + f_K[Y_{j+1}(t)] \\
&\cdot \\
&\cdot \\
&\cdot \\
&m_2\ddot{x}_2(t) + f_P[\dot{Y}_2(t)] + f_K[Y_2(t)] = f_P[\dot{Y}_3(t)] + f_K[Y_3(t)] \\
&m_1\ddot{x}_1(t) + f_P[\dot{Y}_1(t)] + f_K[Y_1(t)] = f_P[\dot{Y}_2(t)] + f_K[Y_2(t)]
\end{aligned} \tag{38}$$

Summing the top j equations at a time with $j = 1, 2, \ldots, N$, yields N equations as follows:

$$f_P[\dot{Y}_j(t)] + f_K[Y_j(t)] = \sum_{k=j}^{N} m_k\ddot{x}_k(t), \quad j = 1, 2, \ldots, N \tag{39}$$

In Eq. 38, because the quantities on the right-hand side are known, the summation of forces can be estimated at any given time.

The response data of two test structures (MF1 and H2) were used by Toussi to evaluate the effectiveness and applicability of the proposed method (126). The MF1 test structure is a one tenth-scale, ten-story, three-bay reinforced concrete structure which was tested by Healey and Sozen (71) at the University of Illinois. The test procedure included a series of strong base motions simulating a scaled-version of the north–

south component of the El-Centro earthquake of 1940. The input acceleration level was magnified for each one of the three test runs. The H2 test structure was also a one tenth-scale ten-story reinforced concrete structure which was tested by Cecen and Sozen (24) and subjected to seven simulated earthquake excitations. The base motions were also taken from the recorded motions of the 1940 El-Centro earthquake.

The removal of the noise trend hidden in the measured acceleration response was accomplished through fitting a polynomial of degree five to each one of the velocity data obtained by integrating the corresponding acceleration time-histories. Figure 10 shows the comparison between the recorded and derived displacements.

Toussi (126) then applied the hysteresis identification method to estimate the inter-story load–deflection relationships of the frame. The spring force is restricted to a polynomial of degree three while the viscous damping is chosen to have a linear form. Schiff (110) stated that, because most large structures exhibit light damping, their identification can be reduced to the finding of the system's lower natural frequencies and modal dampings. His comments seem to be compatible with the assumption of linear viscous damping. Figures 11–13 present the identified behavior of MF1 floors for test runs 1 through 3, respectively. The general softening of the structure is the most apparent feature of these results.

The estimated behavior of the seventh floor of H2 test frame for the seven test runs is shown in Fig. 14. For this frame, the general softening is concluded and as the intensity of the earthquake excitation increases, more nonlinearity in the structural behavior appears. Another interesting feature is the 'soft-to-stiff' type of behavior that the structure experiences under different levels of load. Finally, the identified load–deflection curves become rather wide areawise which is the indication of the dissipation of energy.

4.4 Discussion

During these past two decades, the role of system identification in structural dynamics has been the estimation of certain parameters in differential equations representing the dynamical behavior of structures with known forcing functions and response records. Frequently, the usefulness of such 'realistic' equations of motion was extended beyond the linear or slightly nonlinear range for which the tests were conducted.

Several currently available techniques of system identification are reviewed. In addition, certain methods which are potentially useful in safety evaluation of existing structures are summarized. These methods include the ones by Chen (26) and by Toussi (126). In Chen's methods (26) the

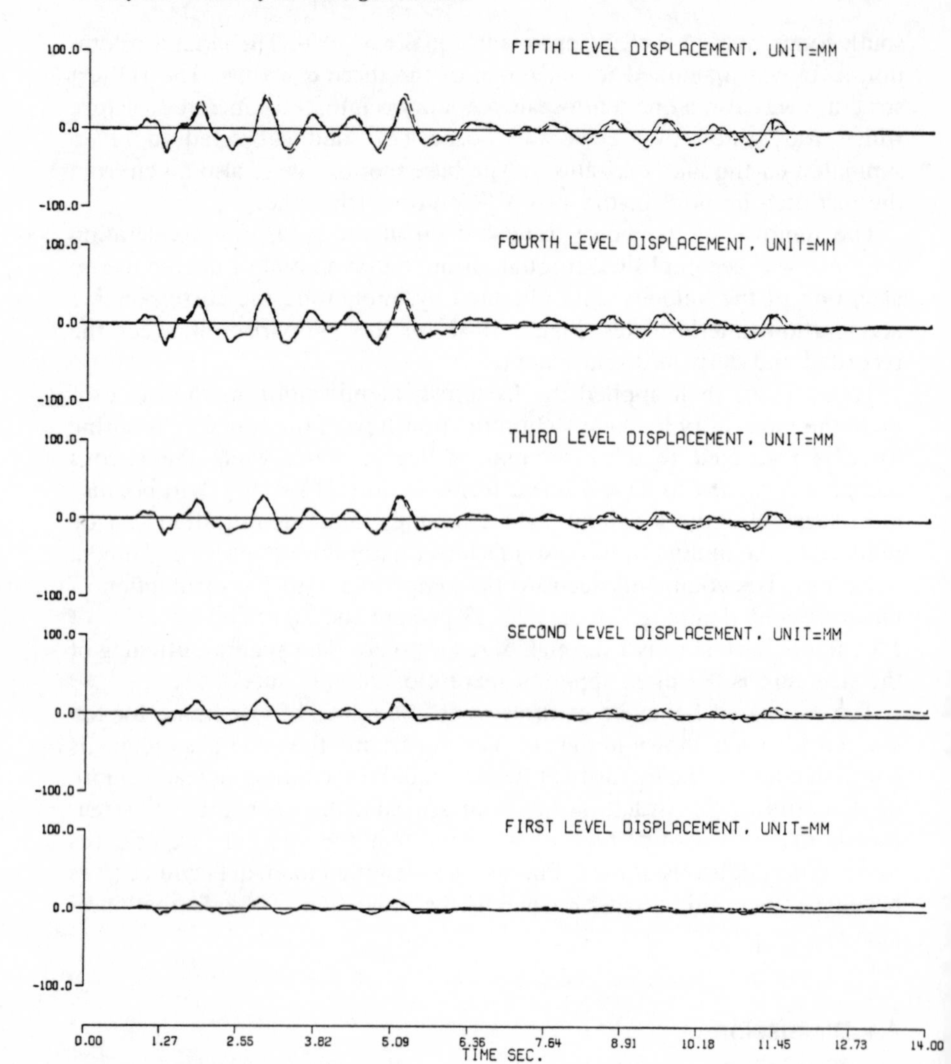

Fig. 10 Comparison between recorded (——) and derived-by-integration (---) displacement for Run 3, MF1

changes in natural frequencies are estimated using linear analyses by not considering the large excursions in structural responses. These detected changes can be used as indicators of moderate structural damage. In Toussi's nonparametric approach (126), hysteretical load–deformation relationships can be estimated. Such estimated hystereses can be useful in the assessment of structural damage.

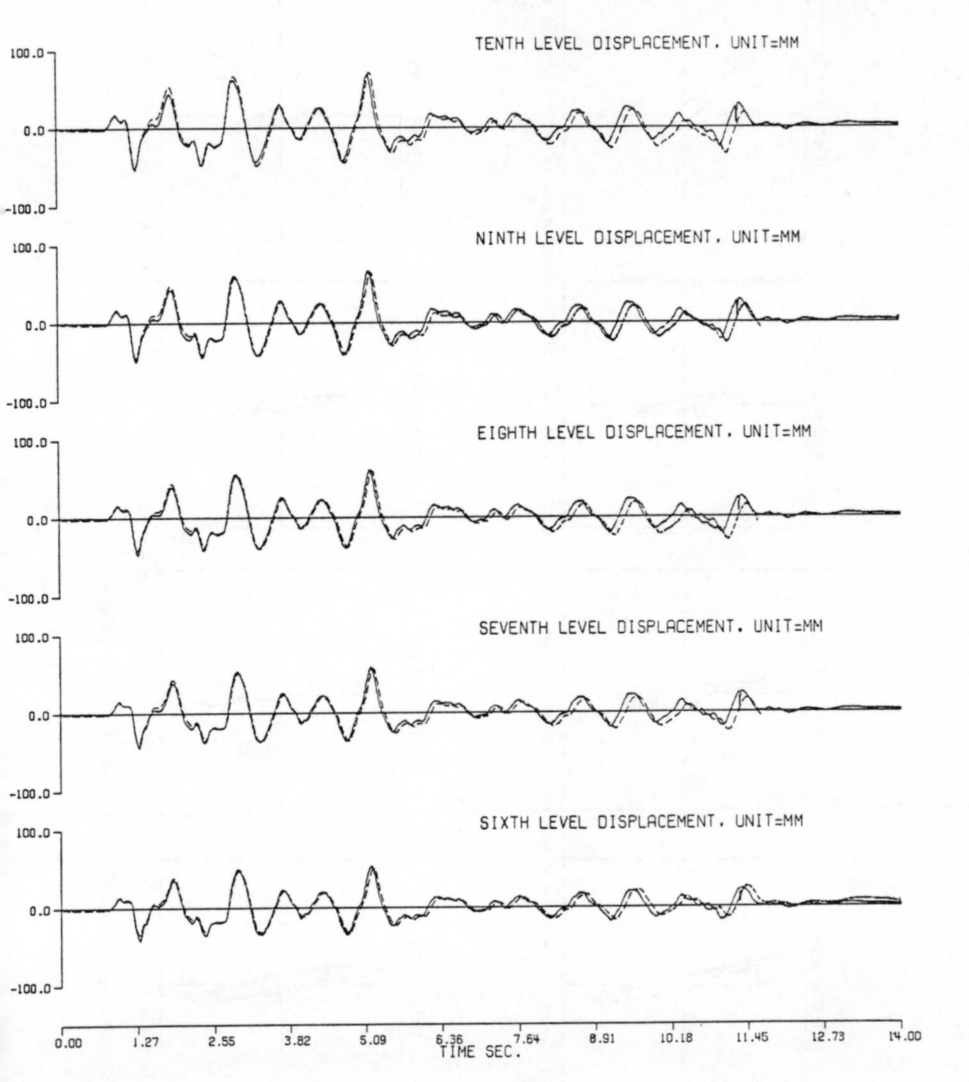

Fig. 10—*contd.*

In general, results of system identification studies can be used as inputs to a decision-making process for safety evaluation of existing structures. The applications of the production system and the theory of evidence are described in Chapter 5.

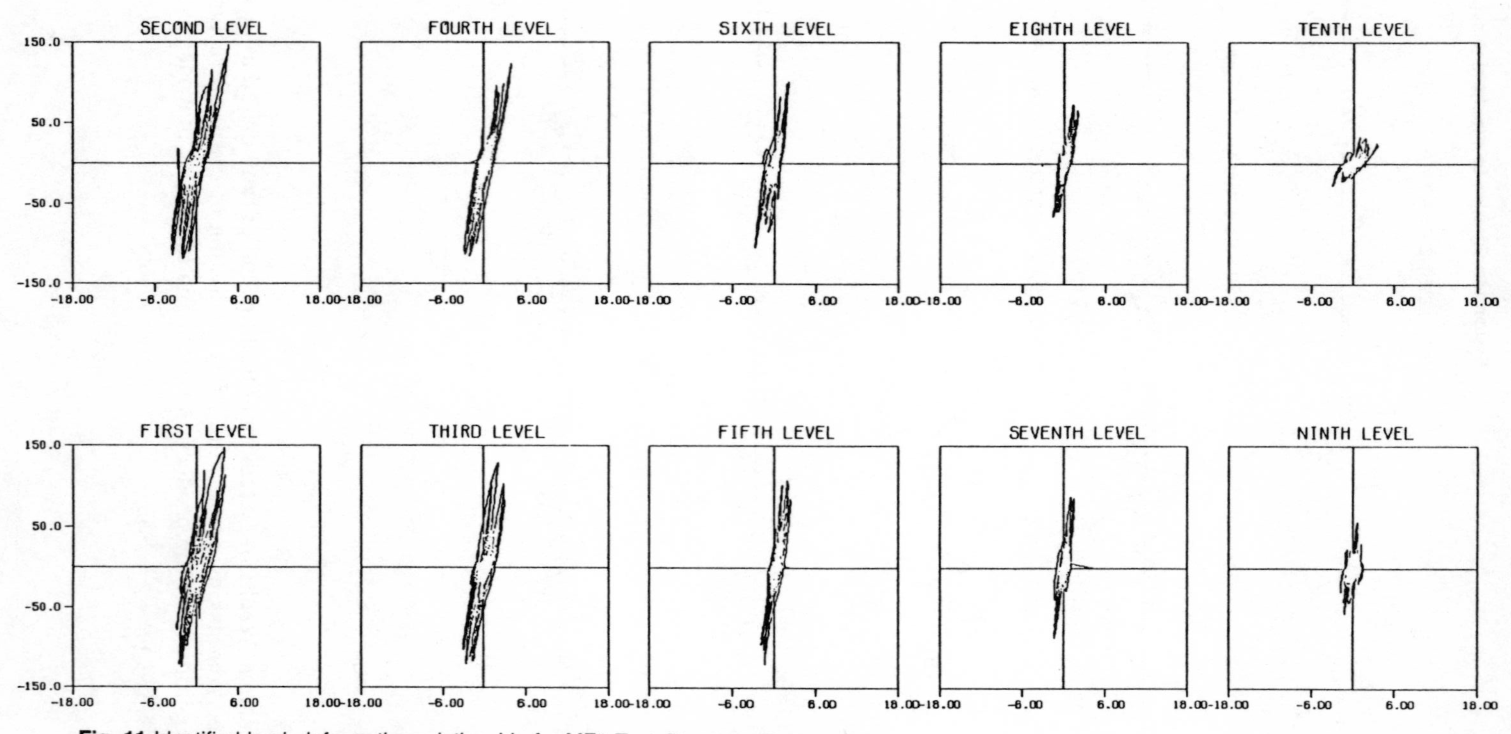

Fig. 11 Identified load–deformation relationship for MF1 Test Structure, Run 1 (126)

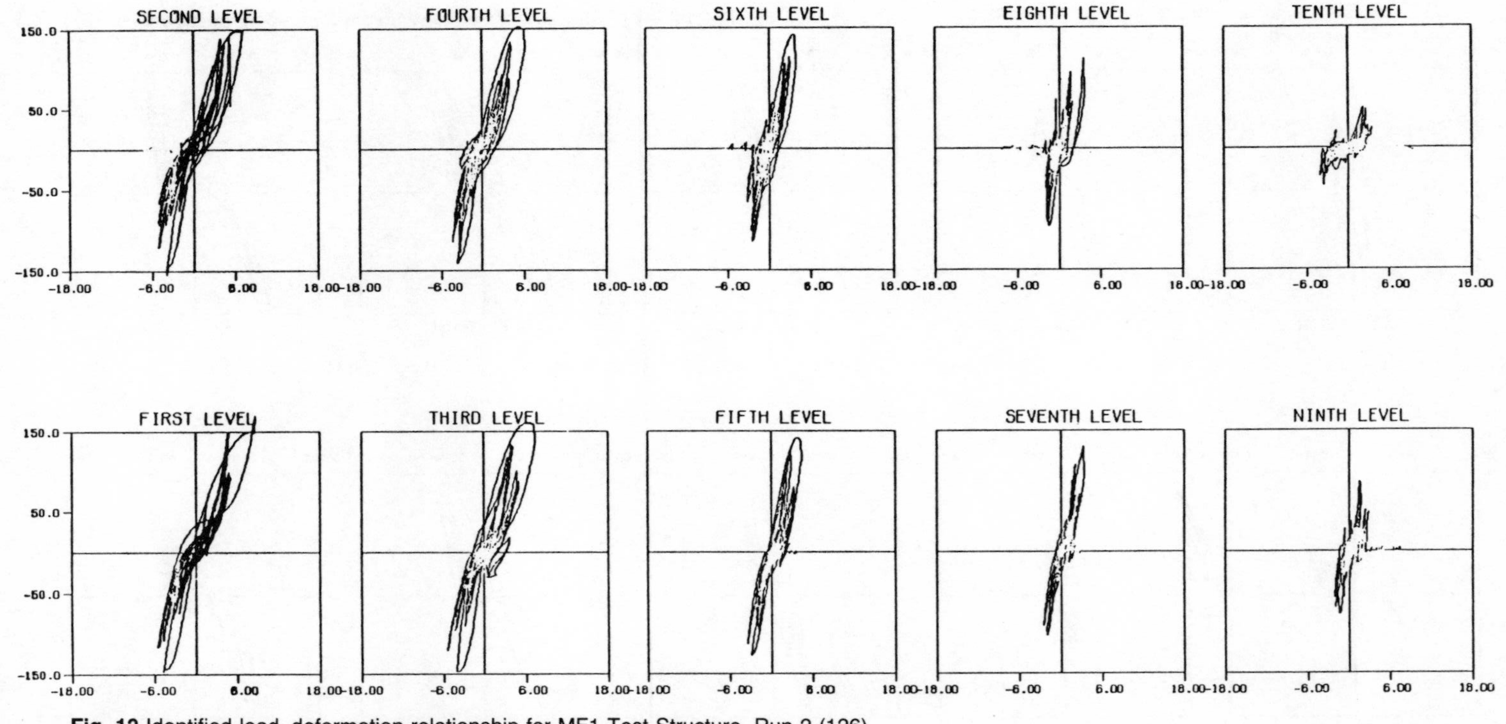

Fig. 12 Identified load–deformation relationship for MF1 Test Structure, Run 2 (126)

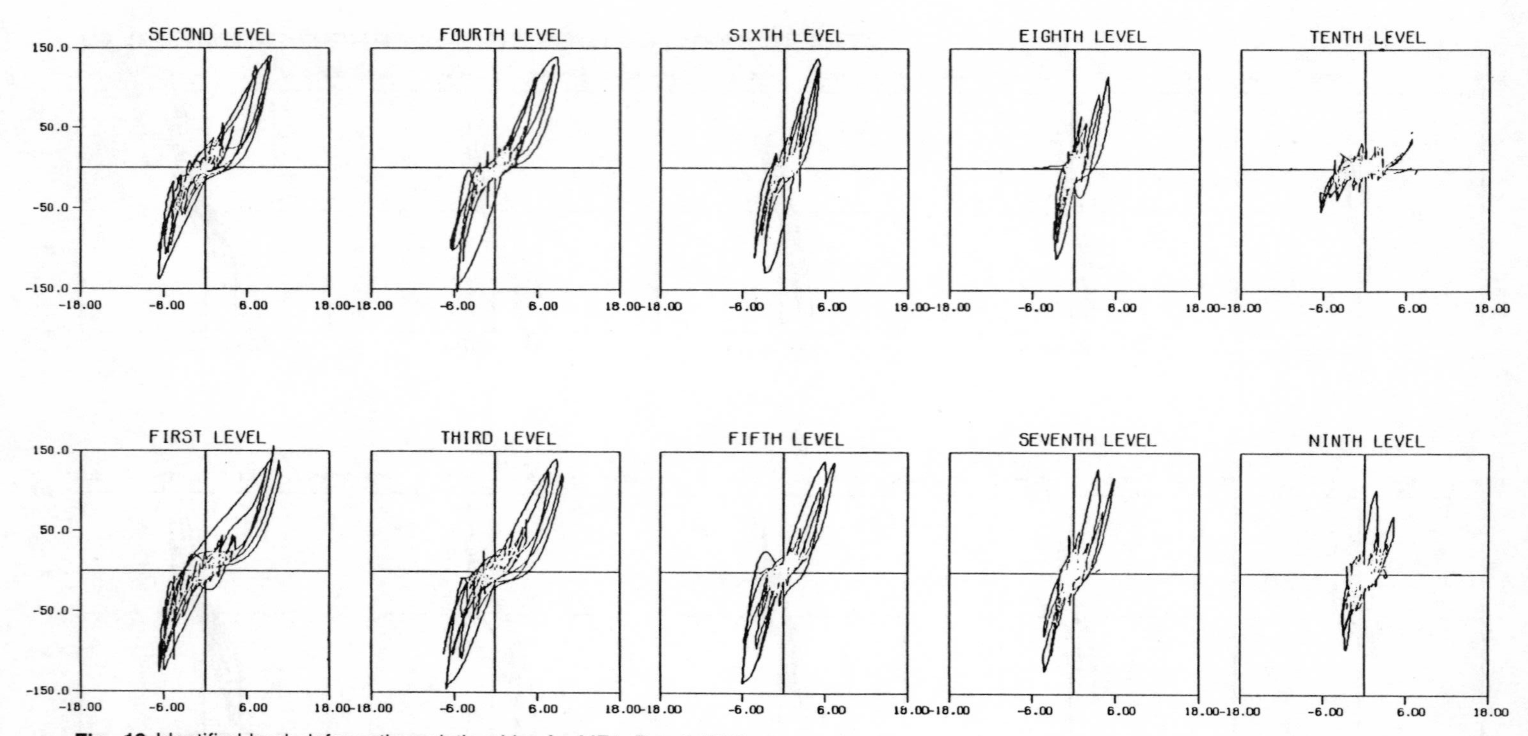

Fig. 13 Identified load–deformation relationships for MF1, Run 3 (126)

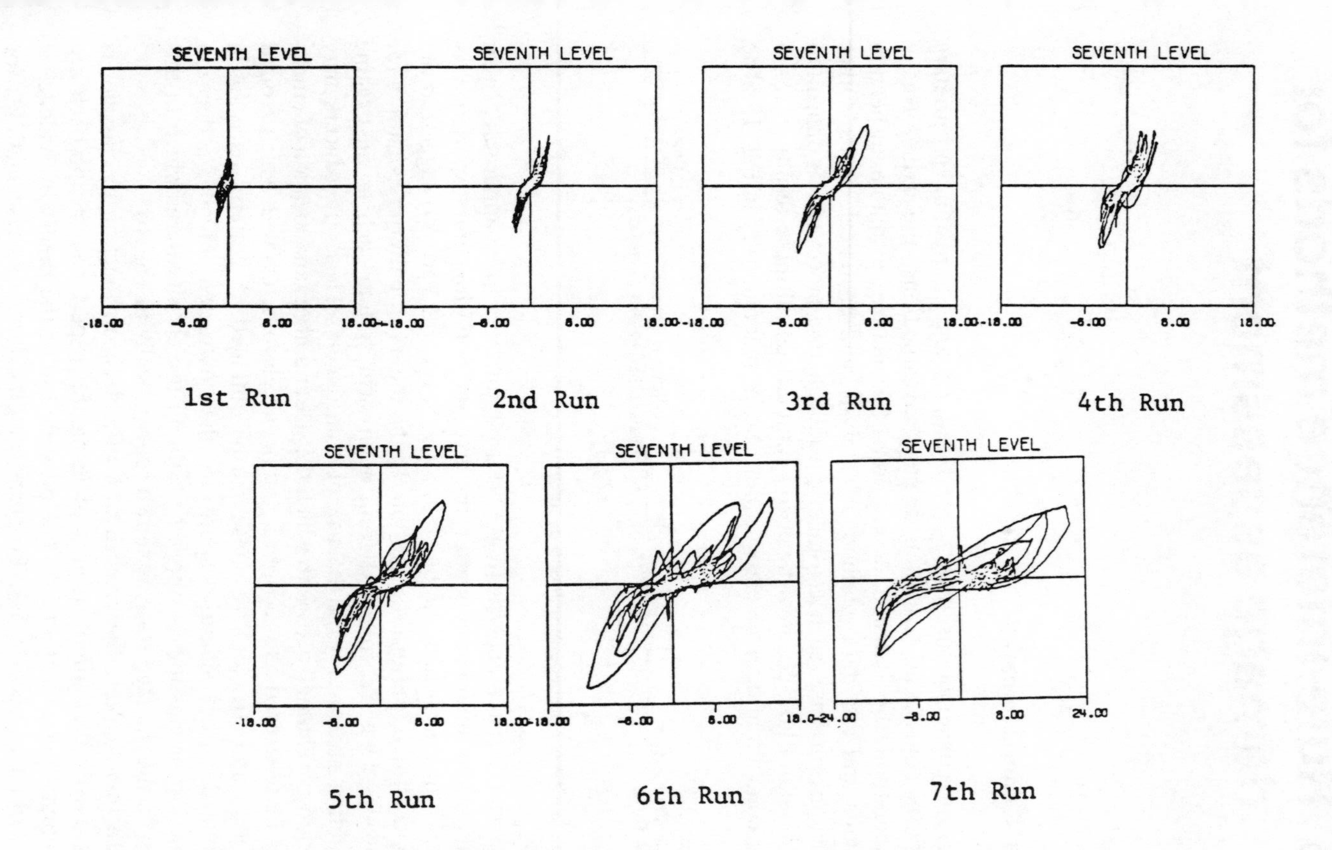

Fig. 14 Comparison between the behaviors of the seventh floor for the seven test runs (H2 Frame) (126)

5 Rule-inference methods for damage assessment

5.1 General remarks

During conversations with several experts and in reading proprietary reports concerning structural damage investigations, the writer usually understood the detailed description of structure, available inspection results and test data, methods of analysis, and analytical results. Other than obvious cases such as total or partial collapse, however, it is difficult to understand how the investigator(s) summarized all these results to reach the concluding damage classification shown schematically in Fig. 15. The

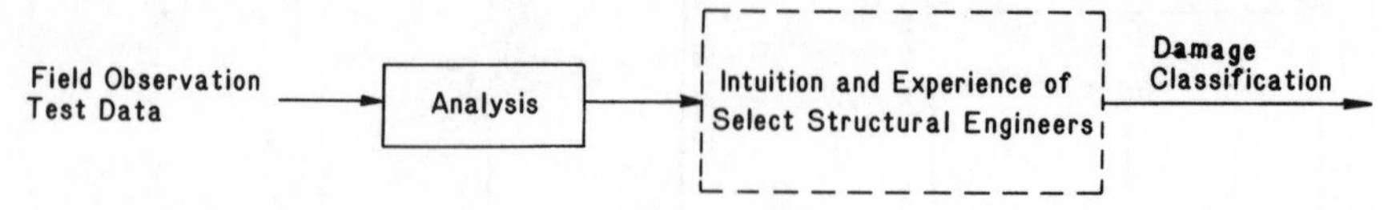

Fig. 15 Current practice of damage assessment

classical decision analysis (e.g., 29) does not seem to be applicable. In an attempt to understand better the experts' intuition and experience as shown in the dashed-line block in Fig. 15, Fu and Yao (59) suggested the application of pattern recognition. In the theory of pattern recognition (3, 56–60), data are collected from a physical system such as an existing building structure with the use of transducers. These transducers may include (*a*) human eyes with which the size, number, and location of cracks can be measured and recorded, and (*b*) accelerometers with which ground motion and structural response can be obtained. A pattern space and a reduced-pattern (feature) space are then extracted. Finally, a decision function or classifier is applied to obtain the classification, which in this case is the damage state as shown schematically in Fig. 16.

In this chapter, the construction of a decision function or classifier is discussed. Preliminary results of using the expert system approach are summarized in Section 5.2. The possible use of the theory of evidence is described in Section 5.3. Proposed modifications of these methods for practical applications are presented in Section 5.4.

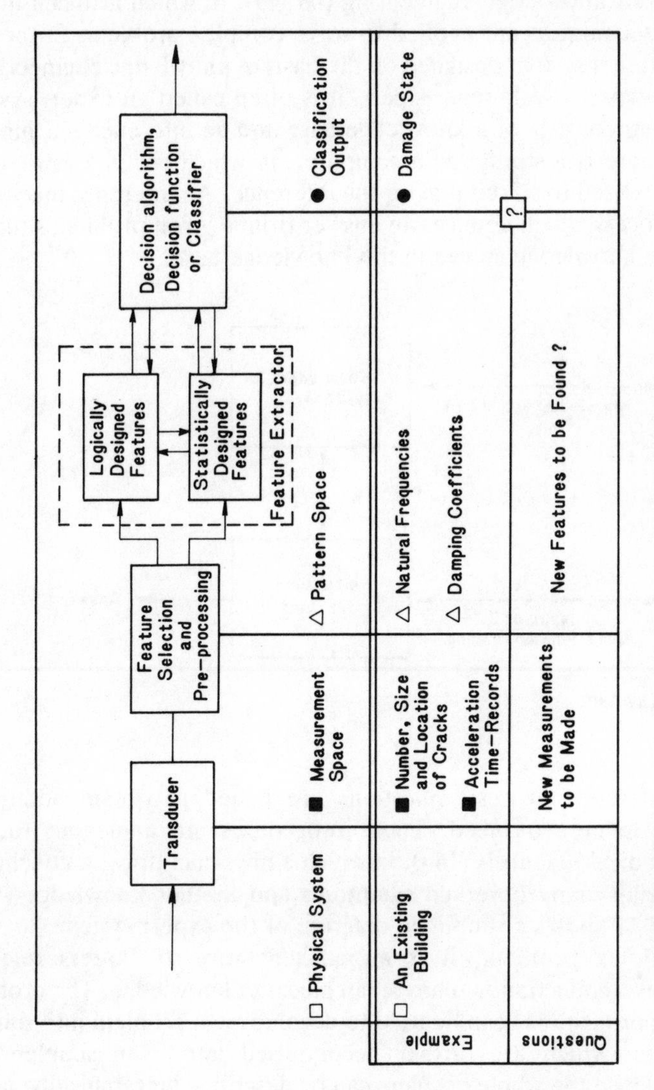

Fig. 16 Application of pattern recognition to damage assessment (59)

5.2 Production system approach

The efficient utilization of human knowledge is a main concern in the subject area of knowledge engineering (48, 49), in which artificial intelligence (AI) techniques are applied to solve complex problems in the real world. Because a system designed on the basis of knowledge engineering is intended to work as a human expert, it is often called an expert system. Such a system consists of a knowledge base and an inference machine. A knowledge base is a storage in a computer, in which useful knowledge is stored in a stylized form suitable for the inference. An inference machine is a control process which deduces an answer from a given problem situation by using the knowledge stored in the knowledge base.

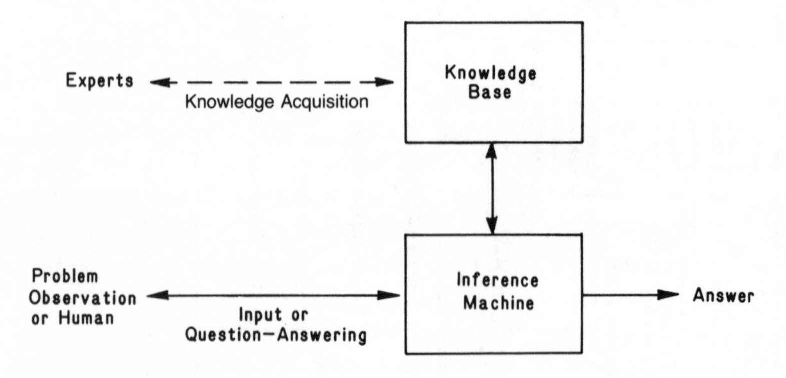

Fig. 17 Expert system

In the inference process, questions are issued to obtain additional information in case of need. Those procedures are analogous to, for example, medical diagnosis (144), in which a physician draws a conclusion by synthesizing many observed symptoms and his/her knowledge (117–119). Figure 17 shows a simplified diagram of the expert system.

In a complex problem, it is an efficient way to express relevant knowledge as a collection of many small pieces of knowledge. The problem reduction approach (81) can be used to decompose a problem into simpler sub-problems, which are further decomposed into even simpler sub-problems. Hence the whole problem can be described hierarchically, and it has its own final goal to be achieved. Likewise each sub-problem has its own sub-goal to be achieved from available information. The production system approach (37) provides a convenient way to express a piece of knowledge for the inference process which infers a higher sub-goal from observed evidences and low sub-goals.

In a production system, a piece of knowledge is written as a production rule (46) in the following basic form:

Rule: IF X,
THEN H,

where IF and THEN clauses are called premise (condition) and action (conclusion), respectively. The function of the rule is that if the premise is satisfied, then the updating action of the sub-goal takes place. If the knowledge is not exact, the rule can be written with a certainty measure C, which may have a value between 0 and 1, as follows:

Rule: IF X,
THEN H with C.

Moreover, as it will be explained later, if the restrictions implied by the premise and action clauses are not precise, they may be expressed in terms of fuzzy subsets (145–147). The fuzzy set theory has been applied to solve civil engineering problems recently (14, 20–23, 45, 114, 141). An introduction to the theory of fuzzy sets is given in Appendix A.

Under this kind of uncertain environment, sub-goals may take intermediate values between those for absolute truth and falsehood. Hence, an inference procedure with uncertainty (80) is required to synthesize available evidences and rules and to assign a rational certainty measure to the sub-goal. With this inference mechanism, eventually the certainty measure of the hypothesis at the final goal can be obtained, which will provide a reasonable answer for decision-making purposes.

Civil engineering structures are commonly classified according to their structural materials into the following types: (*a*) wood, (*b*) masonry, (*c*) reinforced concrete, and (*d*) steel. During construction, certain parts of the structure can be pre-fabricated for economical reasons. In particular, reinforced concrete can be further classified into poured-in-place (or in-situ) and precast and/or prestressed. As a structure with a mixed property of reinforced concrete and steel frame, (*e*) steel-framed reinforced-concrete structures are sometimes built in Japan. Among these types, the emphasis is placed on reinforced concrete and steel structures because they are most frequently used at present.

As the first step, define the damage state of existing structures in terms of a numerical quantity between 0 and 10, where 0 and 10 correspond to 'no damage' and 'total collapse', respectively. In addition, the damage state is defined verbally. Each class is assumed to be associated with a suitable recommendation and the cost for an appropriate repair action.

It is desirable to construct a rational decision-making system for confirming the hypothesis that the structure in question is severely damaged, to be true or false, or to be more reasonable than other hypotheses from possible observations. The observations may come from (i) visual inspection at various portions of the structure, (ii) reading of accelerometer records during the earthquake, (iii) nondestructive testing, and (iv) loading tests before and after the earthquake. Although (i) and (ii) are primarily considered at present, acceptability of other observations will be considered in the design later. Available features for damage classification from the visual inspection may include the detection of deformations and cracks in columns, beams, joints, floors, ceilings, external and internal walls, doors, windows, stairs, nonstructural partitions, utilities, elevators, etc. Features to be derived from the accelerometer records by using system identification techniques may include the change of natural frequency of the building vibration, the change of damping factor, the maximum interstory drift and the total energy absorption and dissipation during the earthquake. In addition, we should consider many other conditions regarding the structures, such as structural material, height or number of stories, areas of floors, shapes, soil condition and foundation, the year that the building was constructed, building use, design parameters if available, existence of walls, experience of human inspector, etc. which are stored and utilized for the inference as reference data apart from inspection data.

The methodology described here allows us to decompose a complex problem into a number of simpler sub-problems. To accommodate knowledge efficiently from human experts, these sub-problems are fitted into knowledge units of the experts. With this in mind, the framework for knowledge representation is determined as shown in Fig. 18, where several intermediate diagnostic states are introduced. Each numbered node corresponds to a set of rules in the production system for inference. For example, Rule 201 is associated with node 2 as follows:

Rule 201	IF :	MAT	is reinforced concrete,
	THEN IF :	STI	is no,
	THEN :	GLO	is no with 0.6,
	ELSE IF :	STI	is slight,
	THEN :	GLO	is slight with 0.6,
	ELSE IF :	STI	is moderate,
	THEN :	GLO	is moderate with 0.6,
	ELSE IF :	STI	is severe,
	THEN :	GLO	is severe with 0.6,
	ELSE IF :	STI	is destructive,
	THEN :	GLO	is destructive with 0.6.

where MAT = material, STI = stiffness, GLO = global damage which is classified into ‘no damage’, ‘slight damage’, ‘moderate damage’, ‘severe damage', and 'destructive damage'. In addition, the membership functions of fuzzy subsets can be assigned as follows:

$$\begin{aligned}\mu_{\text{no}}(\text{d}) &= 1|0 + 0.5|1,\\ \mu_{\text{slight}}(\text{d}) &= 0.5|1 + 1|2 + 0.5|3,\\ \mu_{\text{moderate}}(\text{d}) &= 0.5|3 + 1|4 + 0.7|5 + 0.3|6,\\ \mu_{\text{severe}}(\text{d}) &= 0.3|5 + 0.7|6 + 1|7 + 0.7|8 + 0.3|9,\\ \mu_{\text{destructive}}(\text{d}) &= 0.3|8 + 0.7|9 + 1|10.\end{aligned} \tag{40}$$

For the inference procedure with uncertainty and fuzzy restriction, two statistical methods using Bayesian probability and Dempster and Shafer's

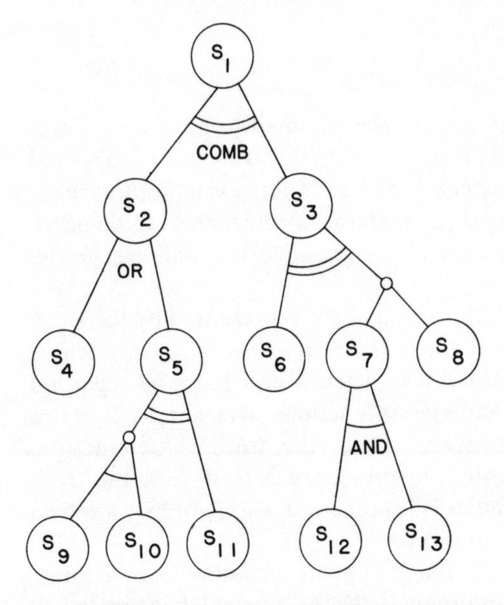

Fig. 18 An example of AND/OR/COMB graph for a problem with uncertainty

probability are described elsewhere (80). The process of fuzzy reasoning as described here is an approximate reasoning process which is compatible with human intuitions. The advantage of fuzzy reasoning as an inference process under the framework of the problem reduction method is that it can yield an approximate answer even when the statistical methods are not applicable. Statistical methods often require idealized assumptions such as the independence of evidences and the mutual exclusiveness and exhaustiveness of hypotheses.

To formulate the inference procedure, consider the basic case that hypotheses in a sub-goal are supported by two lower states X and Y. Their associated rules have been written as,

Rule 1	IF :	X	is X_1
	THEN :	H	is H_{X1} with t_{X1}
	ELSE IF :	X	is X_2
	THEN :	H	is H_{X2} with t_{X2}

.
.
.

Rule 2	IF :	Y	is Y_1
	THEN :	H	is H_{Y1} with t_{Y1}
	ELSE IF :	Y	is Y_2
	THEN :	H	is H_{Y2} with t_{Y2}

.
.
.

where X_l, Y_l, H_{Xl} and H_{Yl} $(l = 1, 2, \ldots)$ are assumed to be fuzzy subsets respectively characterized by $\mu_{X_l}(x)$, $\mu_{Y_l}(y)$, $\mu_{H_{Xl}}(h)$ and $\mu_{H_{Yl}}(h)$, and t_{Xl} and t_{Yl} represent the truth values between 0 and 1. In this example, the fuzzy subsets are no damage, slight damage, moderate damage, severe damage, and destructive damage, and x, y and h are numerical damage grades between 0 and 10.

Suppose that two evidential states X and Y are represented by the truth values of their elements, $\tau(x_i)(i = 1, \ldots, I)$ and $\tau(y_j)(j = 1, \ldots, J)$, which may be the results of preceding inference. Then X and Y can be regarded as fuzzy subsets of which the membership functions $\mu_X(x_i)$ and $\mu_Y(y_j)$ are equal to $\tau(x_i)$ and $\tau(y_j)$, respectively. Thus, the truth value and the membership function are the same. Our objective is to deduce the truth value $\tau(h_k)(k = 1, \ldots, K)$ or membership function $\mu_H(h_k)$ of hypothetical state H from X, Y and the available rules.

For the inference by the conditional statement including fuzzy expression, the compositional rule of inference (146, 147) is widely accepted as fuzzy reasoning or fuzzy logic. This rule is the mapping of the min–max product of a fuzzy subset and a fuzzy relation. There are, however, several choices in the construction of the fuzzy relation (45).

For the rule-based inference with uncertainty, Ishizuka *et al.* (80) constructed fuzzy inference procedure utilizing the compositional rule of inference. It is understood that the minimum (or intersection) aggregation is justified when each expressed fuzzy subset represents the least but dependable creditability of each individual, and the maximum (or union) aggregation is reasonable when each fuzzy subset represents the highest admissible possibility of each individual. For convenience, let us identify the former as credibility and the latter as possibility in this book. The minimum and maximum aggregations are obviously irrelevant to the dependency of the evidences on which the opinions are based; however, the confirmation effect by the combination of different evidences cannot be expected through these aggregations.

According to the truth qualification technique of fuzzy subset (147), let the fuzzy subset H_{Xl} (H_{Yl}) of the THEN clause be modified by the accompanied truth value t_{Xl} (t_{Yl}) to H'_{Xl} (H'_{Yl}). Depending on the properties of the credibility and possibility, define for credibility,

$$\mu_{H'_{Xl}}(h) = t_{Xl} \cdot \mu_{H_{Xl}}(h), \tag{41}$$

and for possibility,

$$\mu_{H'_{Xl}}(h) = t_{Xl} \cdot \mu_{H_{Xl}}(h) + (1 - t_{Xl}). \tag{42}$$

So far as truck qualification is concerned, these modifications are equivalent to the qualification with the use of the linguistic truth values. The effect of the modification by t_{Xl} is the decrease of the positive degree and the negative degree from 1 of the fuzzy subset H_{Xl} for credibility and possibility, respectively.

When constructing a fuzzy relation R_{Xl} from the statement of 'IF X is X_l, THEN H is H'_{Xl}', it seems to be reasonable to select the tightest relation for credibility and the loosest relation for possibility among several choices (45) as follows,

$$\mu_{R_{Xl}}(x, h) = \min\{\mu_{X_l}(x),\ \mu_{H'_{Xl}}(h)\}, \text{ for credibility} \tag{43}$$

and

$$\mu_{R_{Xl}}(x, h) = \min\{1,\ 1 - \mu_{X_l}(x) + \mu_{H'_{Xl}}(h)\}, \text{ for possibility.} \tag{44}$$

Then the composite fuzzy relation from Rule 1 is obtained as the union or intersection of the individual fuzzy relations as,

$$R_x = \bigcup_l R_{X_l}, \text{ for credibility} \tag{45}$$

$$R_x = \bigcap_l R_{X_l}, \text{ for possibility} \tag{46}$$

where $\cup$ and $\cap$ denote union and intersection, respectively.

The min–max product gives us an approximate answer from the fuzzy state X and the fuzzy relation R_x between the domains of X and H as follows,

$$H_x = X \circ R_x, \tag{47}$$

which can be applied to both credibility and possibility and means that,

$$\mu_{H_x}(h) = \max_x \min\{\mu_X(x),\ \mu_{R_x}(x, h)\}. \tag{48}$$

Similarly, we can have approximate inferred result H_Y from the fuzzy state Y and Rule 2.

As described before, H_X and H_Y are to be aggregated as,

$$H_{XY} = H_X \cup H_Y, \text{ for credibility} \tag{49}$$

and

$$H_{XY} = H_X \cap H_Y, \text{ for possibility} \tag{50}$$

or

$$\mu_{H_{XY}}(h_k) = \max \{\mu_{H_X}(h_k), \mu_{H_Y}(h_k)\}, \text{ for credibility} \tag{51}$$

and

$$\mu_{H_{XY}}(h_k) = \min \{\mu_{H_X}(h_k), \mu_{H_Y}(h_k)\}, \text{ for possibility.} \tag{52}$$

In short, the fuzzy reasoning procedure as given by Eqs. 41, 43, 45, 47, and 49 yields the credibility whereas the procedure as given by Eqs. 42, 44, 46, 48, and 50 yields the possibility. Although the inference process from two evidential states has been described, the supports from more than two evidential states can be combined by using Eqs. 49 or 50.

Because the fuzzy reasoning itself has no confirmation effect, it is desirable for a better result to generate combined rules and to add them to the knowledge base. With these inference procedures, the truth value propagates through the inference network and eventually the truth value of the hypothesis at the final goal is determined. The meaning of truth value or the degree of membership is rather vague in the physical sense, though it can be said that the higher the value, the more credible or possible the fact is. Note that the possibility is always larger than the credibility as defined here.

A preliminary version of such a system called SPERIL-I has been developed for the purposes of illustration (*see* Appendix B). At present, it is being modified and expanded for more practical applications.

5.3 Theory of evidence and its application to damage assessment

The likelihood of proposition A is expressed with a subinterval, $(s(A), p(A))$, of the unit interval, $(0, 1)$. The lower value, $s(a)$, represents the 'support' for the proposition, while the upper value, $p(A)$, indicates the degree to which one fails to doubt A (38, 113). 'Support' is considered as the total positive effect a body of evidence has on a proposition, while $p(A)$ represents the total extent to which a body of evidence fails to refute a proposition. The degree of uncertainty of A is the difference between the upper and the lower probability values (i.e., $p(A) - s(A)$). For instance, if no information is available about A, the Shafer's representation becomes $A_{(0,1)}$; while $A_{(1,1)}$ and $A_{(0,0)}$ indicate that A is true and false, respectively. A partial support for A might be indicated by $A_{(0.30,0.85)}$ while $A_{(0.30,1)}$ expresses a firm support for A.

The frame of discernment, Θ, is a set whose subsets are propositions. When a proposition corresponds to a subset of a frame of discernment, it means that the frame discerns that proposition. This is most effective when one is concerned with the true value of a quantity. If the quantity is indicated by θ and the set of its possible values by Θ, then the propositions of interest are precisely those of the form 'the true value of θ is in T', where T is a subset of Θ. It should be noted that the 'possibilities' that comprise Θ will get meaning from what we know or think we know and it is not independent of our knowledge (113). The primary advantage of this formation is that it translates the logical notions of conjunction, disjunction, implication, and negation into the set-theoretic notions of intersection, union, inclusion and complementation.

Dempster (38) introduced a technique to combine the information as supplied by different knowledge sources. A belief function which is constructed to represent an evidence can be combined with another belief function, representing another evidence, in order to pool the information obtained from these two knowledge sources. Dempster's rule deals symmetrically with the two knowledge sources and does not depend upon the priority of either one of them. The combination of two belief functions is called 'orthogonal sum'.

Assume that the jth knowledge source, K_j, provides evidential information about a set of propositions. This information can be presented by the following belief function:

$$\begin{aligned} & m_j : \{A_i \mid A_i \subset \Theta\} \rightarrow [0,1] \\ & m_j(\phi) = 0 \\ & \sum_i m_j(A_i) = 1 \end{aligned} \tag{53}$$

where A_i are subsets of Θ (i.e., $A_i\Theta$). The 'basic probability mass', $m_j(A_i)$, represents a measure of the belief that K_j has committed exactly to proposition A_i. Therefore, m_j may be visualized as a partitioned unit line segment, and the length of each subsegment corresponds to the mass contributed to that subsegment by K_j. The mass assigned to Θ is assumed to be distributed in some manner among the propositions discerned by Θ. In fact, $m_j(\Theta)$ represents the residual 'uncertainty' of the K_j directly. Once the masses assigned to the propositions are obtained, the evidential intervals, $(s(A_i),p(A_i))$ can be determined directly.

The total support for proposition A_i is given by the sum of the masses assigned to A_i and to the subsets of A_i:

$$S_j(A_i) = \sum_k m_j(D_k) \tag{54}$$

where $D_k \subset A_i$. The plausibility of A_i, $P(A_i)$, is, on the other hand, one minus the sum of the masses assigned to A_i and to the subsets of A_i:

$$P_j(A_i) = 1 - \sum_k m_j(B_k) \tag{55}$$

where $B_k \subset A_i$. Consequently the uncertainty of A_i is obtained by:

$$U_j(A_i) = P_j(A_i) - S_j(A_i) \tag{56}$$

By repeating the rule of combination, any number of belief functions can be combined. The process is given as follows:

$$\begin{aligned}&\mathrm{Bel}_1 + \mathrm{Bel}_2\\&(\mathrm{Bel}_1 + \mathrm{Bel}_2) + \mathrm{Bel}_3\\&((\mathrm{Bel}_1 + \mathrm{Bel}_2) + \mathrm{Bel}_3) + \mathrm{Bel}_4\\&\text{etc.}\end{aligned}$$

It is continued until all the Bel_i are included. It is to be noted again that the Dempster's rule of combination is independent of the priority of the knowledge sources.

Toussi *et al.* (126, 129) applied the theory of evidence to determine the damage state of an existing structure, their frame of discernment being the set of possible states of damage, {S, L, D, C}, where S, L, D, and C denote 'Safe', 'Lightly damaged', 'Damaged', and 'Critically damaged', respectively. Two features, namely 'slope ratio' and 'drift ratio', are considered as the damage state parameters. The information about the parameter values is obtained in the form of parameter mass distribution graphs. These curves indicate the probability of any state of damage for a given parameter value. To develop a damage parameter mass distribution, information regarding the damage state of a prototype structural frame, which had gone under a series of successive dynamic tests is used. Then the information about the states of damage of another dynamically tested structure is used to evaluate the developed parameter mass distributions. It should be noted, in the foregoing sections, that the terminologies 'training samples' and 'testing samples' refer to the information used to develop and to test the parameter mass distributions, respectively.

Cecen (24), using the shaking table of the University of Illinois, recorded the response of a one-tenth scale and ten-story structural frame subjected to a series of simulated earthquakes. The simulated earthquakes were patterned after the north–south component of the acceleration history as recorded in El-Centro during the Imperial Valley earthquake of 1940. The test structure was subjected to seven successive earthquake test runs. The base motion intensities were incrementally increased with each successive run. Toussi and Yao (127, 128) used the measured acceleration response histories to identify the inter-story hysteretic behavior of the test structure.

The existence of a 'soft-to-stiff' property in the structure's behavior becomes more apparent as the structure experiences larger deflections. This property is indicated by the slope ratios calculated for each floor at different test runs. The drift ratio is used as another measure of damage (120). The classification of the damage state of a floor of the test structure H2 results from three different pieces of information: comments and reports made by Cecen (24) who tested and analyzed the measured response of H2, the crack patterns, and the hysteretic behavior of floors as identified by Toussi and Yao (126–128).

As testing samples, data from the MF1 test structure—which was a one-tenth scale, ten-story reinforced concrete frame dynamically tested at the University of Illinois—were used (71). The dynamic test procedure included a series of strong base motions, simulating a selected version of the north–south component of the El-Centro earthquake of 1940. To evaluate the proposed algorithm, a similar procedure which was used for the classification of H2-frame floors' damage states is also conducted for the MF1 structure. Then a comparison between the results of this classification and those of the proposed algorithm is made to find how effective and accurate the algorithm is.

Results indicate that five out of twenty-five cases of damage assessments made by using the proposed algorithm are likely to be incorrect. A twenty per cent error is acceptable considering the fact that the lack of sufficient samples did not permit an effective statistical analysis for the determination of the damage parameter mass distribution to be made. Moreover, only two features are used here. In practice, there are usually other features available for damage assessment of existing structures.

5.4 Possible improvements

In the classical theory of structural reliability (4, 53–55), the theory of probability and statistics is applied to interpret the degree of safety in terms of reliability (probability of survival) or failure probability for a given type of structure. For such computations, it is necessary to have (*a*) sufficient statistical data concerning the loads and resistances of these structures, and (*b*) adequate knowledge of the structural behavior under various loading conditions and failure paths. For many existing structures such as buildings and bridges, the design and construction of each structure are unique. Because of the massive size and high cost involved, it is usually not feasible to conduct destructive tests of the whole structure even once to gain some understanding of the structural behavior near the 'failure' state(s). Therefore, it is difficult to expect that sufficient statistical data are available for such purposes. Nevertheless, much progress has been made during these past four decades for the estimation of the order of magnitude of failure probabilities using the objective type of data and state-of-the-art methods of structural analysis (74, 75, 86, 98, 103, 111, 115, 116, 122–124, 136–138).

Following Blockley (14) and Brown (20), such an objective safety measure, $n_0 = -\log p_f$, can be modified with subjective input to obtain the fuzzified safety measure N_0 as shown in Fig. 19. With the additional information resulting from ith inspection, this fuzzified safety measure n_i

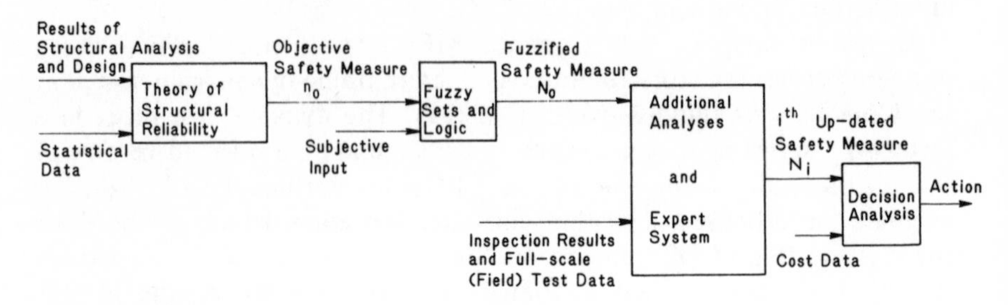

Fig. 19 Flowchart for safety evaluation of structures

can be further modified with additional analyses such as the application of the theory of evidence (113, 126, 129) and an appropriate expert system (80, 81).

The ith revision of the structural reliability as given in Eq. 12 can be rewritten as follows:

$$L_T^{(i)}(t) = (1 - 10^{-N_i})\exp\left[-\int_{t_i}^{t} h_T^{(i)}(t)\mathrm{d}t\right] \tag{57}$$

The estimation of the hazard function $h_T^{(i)}(t)$ is more complicated and requires further investigation (89, 103, 122–124, 136–138).

To help interpret such results, let

$$L_T^{(i)}(t) = 1 - 10^{-N(t)} \tag{58}$$

or

$$N = N(t) = \log_{10}(1 - L_T^{(i)}(t)) \tag{59}$$

where $t - t_i$ denote the time period between ith inspection and expected useful lifetime of this particular structure. Depending on the importance of the structure, the structure may be classified as satisfactory, questionable, or unsatisfactory as shown in Fig. 20. Several possible membership functions for such classifications are given in Fig. 21.

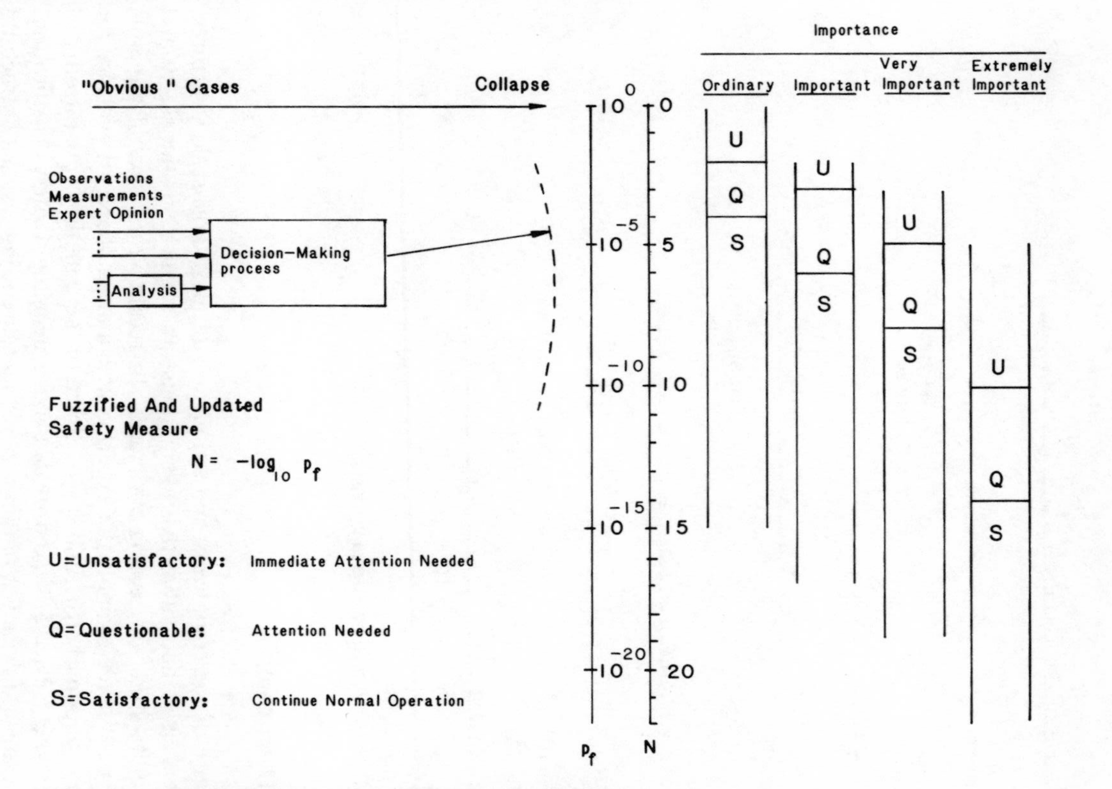

Fig. 20 Suggested reliability and performance classification

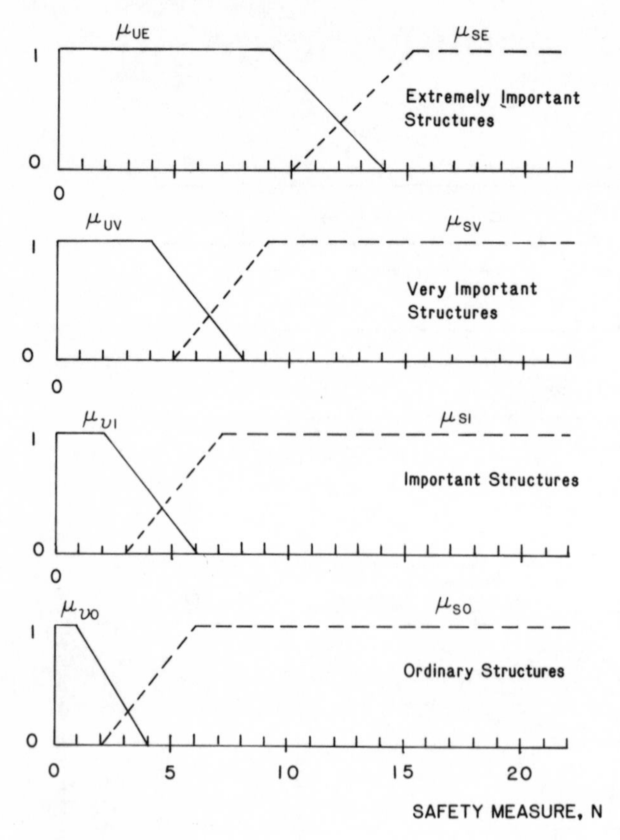

Fig. 21 Possible membership functions

5.5 Discussion

Because of the complexity and uniqueness of most large-scale structures, the expertise and subjective evaluations are required in the safety and reliability study of existing structures. In recent years, several attempts have been made to apply the expert system and the theory of evidence for the evaluation of structural safety. Some of these methods are introduced in Sections 5.2 and 5.3. To provide a more meaningful safety classification, the reader is advised to obtain a safety measure using results of structural analysis and design, statistical data, subject input, and inspection results and test data, with the application of fuzzy sets and an expert system as shown in the flowchart in Fig. 19. To make these improvements, a number of research topics are given and discussed in the next chapter.

6 Summary and concluding remarks

6.1 Summary

Even a new structure may not be completely safe because it can be defective due to errors in analysis and design, fabrication, and/or construction. Consequently, any existing structure may have an initial value of 'damage' in the broad sense. Certainly, there can be additional damage arising from unexpected usage of the structures, wear and corrosion, the occurrence of hazardous events, fatigue and other cumulative damage.

It is desirable to find a generally acceptable and meaningful definition of structural damage for various types of structures. With such a measure of structural damage, the safety of a specific existing structure can be assessed and appropriate decisions regarding repair and maintenance can be made accordingly. Generally, there are three types of definitions for structural damage. The first one is numerical, the second one is given in terms of repair or replacement costs, and the third one is verbal. Frequently, numerical values are also assigned to various verbal classifications. Some of these definitions are reviewed and discussed in Chapter 2. In this book, the term 'damage' refers to any deficiency and/or deterioration of strength as caused by external loading and environmental conditions as well as human errors in design and construction. Therefore, a poorly designed and/or poorly constructed structure can have an initial 'damage' measure while it is still new without experiencing any severe loading conditions.

Engineers usually like to manipulate numbers. Therefore, those definitions of structural damage with numerical values appeal to many engineers. However, it appears that the most numerical scales to date are rather arbitrary. Those definitions involving repair or replacement costs are attractive on the surface. Nevertheless, questions remain as to the process with which such costs can be determined rationally and accurately without any precise knowledge concerning the extent of structural damage. The verbal classifications are meaningful, especially if and when such classifications are made by highly qualified experts. Nevertheless, there exist cases when numerical values are needed for further analysis of structural damage. Following Blockley (14) and Brown (20), a safety measure, N, is defined as the negative logarithm (with base 10) of the probability of failure. The most important advantage for such a safety measure is that it is directly related to the probability of failure in a meaningful manner. Furthermore, it can be determined both objectively (4, 53–55) and subjectively (14, 20). Such a safety measure can be related to verbal classifications in a meaningful way (141).

The current approaches in engineering practice are summarized and reviewed in Chapter 3. As far as the writer understands these procedures, the present decision-making process depends primarily on the experience, intuition, and judgment of these practicing experts. Several proposed methods are also discussed in terms of their potential applications.

Because most existing structures are extremely complex and unique in their design and construction, the mathematical representations of the actual structural behavior are usually unknown (139, 140, 144). The application of various system identification techniques to obtain 'more realistic' mathematical models is reviewed in Chapter 4. The paradox between small-amplitude test data and large-amplitude analyses in such applications is discussed in some detail. Whenever records of structural response to earthquake excitation are available, the potential application of non-parametric methods for the analysis of such records is explored.

To understand how experts make their decisions concerning safety evaluations, applications of rule-inference methods (80, 81, 117–119), the theory of evidence (38, 113, 126, 129), the theory of fuzzy sets (145–147) are summarized and discussed in Chapter 5. The basic theory of fuzzy sets is given in *Appendix* A. A preliminary version of the rule-based system for the evaluation of damage assessment called SPERIL-I is introduced and given in *Appendix* B. Such a system can be useful in training young engineers and assisting senior engineers in making their decisions. Moreover, it will be useful in building up the knowledge base with inputs from many experts who are willing to share their experience with other structural engineers. However, more studies are required for the practical implementation of a rule-based system such as SPERIL-I. Several research topics are described in the next section.

6.2 Concluding remarks

Because of the complexity and uniqueness of most existing structures, there does not exist any 'exact' mathematical model for the description of the structural failure behavior even under a set of prescribed loading and environmental conditions. Consequently, there is still much 'art' in the practice of safety evaluation of existing structures. In current practice, engineers collect inspection results and test data and analyze them in a scientific and objective manner. In summarizing and interpreting these results, which can still be voluminous, however, they still rely on their individual experience, intuition, and judgment.

At present, expert systems, fuzzy sets and the theory of evidence seem to be useful in an attempt to understand better how these experts make their final decisions in the safety evaluation of existing structures. The writer believes that further studies along these lines are promising and should be continued. Specifically, the following topics should be investigated:

(*a*) The safety classification should be linked to the failure probability through the use of the safety measure as introduced by Brown (20) and others. Therefore, one possible improvement for the rule-based system is to replace the arbitrary damage scale with this safety measure.

(*b*) The ith hazard function $h_T^{(i)}(t)$, $t_i \leq \tau \leq t$, represents the risk of the structure between the time for ith inspection, t_i, and expected life of the structure, t, and is a function of the current condition of the structure and the expected loading and environmental conditions during the time interval (t_i, t). It is desirable to estimate the mathematical form of such a function and values of its parameters on the basis of analytical results, available inspection and test data, and expert opinion.

(*c*) To provide additional and meaningful input to the expert system, it is desirable to continue various research activities in the subject area of system identification. In particular, it will be helpful to develop new methods of analysis of response data and/or develop new techniques of testing and recording data which are useful for such purposes.

Although no final answer to the problem of safety evaluation of existing structures is given in this book, several aspects of this subject area are reviewed and discussed in a comprehensive manner. While there are satisfactory solutions in engineering practice, these solutions are not available to young and inexperienced engineers. Moreover, there does not seem to exist any rational basis for comparing results of different experts at present. In recent years, an attempt has been made to formulate a general framework for such purposes. Such a formulation and several suggestions for its practical implementation are summarized here.

Appendix A Elements of fuzzy sets

A1 General remarks

Zadeh (145, 146) stated that our ability to make both precise and significant statements concerning a given system diminishes with increasing system complexity. Consequently, the closer one looks at a real-world problem which is usually complex, the fuzzier its solution becomes. Although the theory of fuzzy sets is relatively new, the calculus of fuzzy sets is well developed with various applications (14, 20–23, 147). The application of fuzzy sets to several civil engineering problems has been reviewed recently (23). In the following, fundamental elements of the theory of fuzzy sets as given by Zadeh (146) and Kaufmann (84) are summarized along with several structural engineering examples from Yao (141) and a simplified version of an example on structural reliability from Brown (20).

The values of a linguistic variable are words, phrases, or sentences in a given language. For example, structural damage can be considered as a linguistic variable with values such as 'severely damaged', 'moderately damaged', which are meaningful classifications but not clearly defined. In many situations, a complex problem can be divided into simpler questions. Some of these questions can best be answered by experienced engineers with descriptive words such as 'large' or 'medium', which are values of a given linguistic variable. The theory of fuzzy sets can be used to interpret such adjectives with membership functions, which can be manipulated in a logical manner to obtain an answer to the original and complex problem.

A2 Basic definitions

A fuzzy set A in a given sample space Ω is a set of ordered pairs $\{\mu_A(x)|x)\}$ for each $x \in \Omega$, where $\mu_A(x)$ is called the membership function and Ω denotes the sample space. If the membership set consists of only two elements, say 0 and 1, then A is said to be an ordinary (or nonfuzzy, or crisp) set. For fuzzy sets, the membership set usually consists of the continuous interval 0 to 1. As an example, let N be the set of natural numbers, i.e., $N = \{0, 1, 2, \ldots\}$. Consider the fuzzy set A of 'small' natural numbers as follows:

$$A = \{(1|0), (0.8|1), (0.6|2), (0.3|3), (0|4), \ldots\} \qquad \text{(A-1)}$$

In words, we say that the number '0' has a nonfuzzy membership, '1' has a 'strong' membership, '2' has a 'fairly strong' membership, '3' has a 'weak' membership, '4' and higher numbers have non-memberships of the fuzzy set A of 'small' natural numbers.

As another example, let y denote the proportion of cracked width of an uniaxially-loaded plate, and let B denote the 'severely damaged' state of this plate. Then, we may write

$$B = \{(0|y < 0.1), (2(y - 0.1)|0.1 \leqslant y < 0.6), (1|y > 0.6)\} \qquad \text{(A-2)}$$

or

$$\mu_B(y) = \begin{cases} 0, & y < 0.1 \\ 2(y - 0.1), & 0.1 \leqslant y \leqslant 0.6 \\ 1, & y > 0.6 \end{cases} \qquad \text{(A-3)}$$

Such a description can be the result of compiling and analyzing the subjective evaluation of a number of experts. As it is given in this contrived example, this plate specimen is clearly (nonfuzzy) severely damaged whenever $y > 0.6$, i.e., the crack length exceeds six-tenths of the width of the plate. On the other hand, the plate specimen is not considered to be in a severely damaged state when $y < 0.1$. In the range $0.1 \leqslant y \leqslant 0.6$, there exists a fuzziness about the definition of the 'severely damaged' state, which is reflected by the linear membership function in this case.

The complement of a fuzzy set A is denoted by $\bar{A}$, and is given by

$$\bar{A} = \{(\mu_{\bar{A}}(x)|x)\} \qquad \text{(A-4)}$$

where

$$\mu_{\bar{A}}(x) = 1 - \mu_A(x) \qquad \text{(A-5)}$$

The membership function for the intersection of two fuzzy sets, say A and B, is given as follows:

$$\mu_{A \cap B}(x) = \min\,(\mu_A(x), \mu_B(x)) \qquad \text{(A-6)}$$

On the other hand, the membership function for the union of two fuzzy sets, say A and B, is as follows:

$$\mu_{A \cup B}(x) = \max\,(\mu_A(x), \mu_B(x)) \qquad \text{(A-7)}$$

The algebraic sum of two fuzzy sets, say A and B, is denoted by $A + B$ and has the following membership function:

$$\mu_{A+B}(x) = \mu_A(x) + \mu_B(x) - \mu_A(x)\mu_B(x) \qquad \text{(A-8)}$$

or, for given values of x,

$$\mu_{A+B} = \mu_A + \mu_B - \mu_A \cdot \mu_B \tag{A-9}$$

More generally, we have

$$\sum_{i=1}^{\mu_n} A_i = 1 - \prod_{i=1}^{n}[1 - \mu_{A_i}] \tag{A-10}$$

The membership of the algebraic product of two sets, say A and B, is given by

$$\mu_{AB}(x) = \mu_A(x)\mu_B(x) \tag{A-11}$$

Therefore, for $\alpha > 0$,

$$\mu_{A^\alpha}(x) = [\mu_A(x)]^\alpha \tag{A-12}$$

As an example, consider an axially-loaded plate in which n cracks (with length C_i, $i = 1, \ldots, n$,) have been detected. Let D_i denote the severe-damage state of this plate due to the ith crack, $y_i = C_i/b$, and

$$\mu_{D_i} = \begin{cases} 0, & y_i < 0.1 \\ 2(y_i - 0.1), & 0.1 \leqslant y_i \leqslant 0.6 \\ 1, & y_i > 0.6 \end{cases} \tag{A-13}$$

Let B denote the overall damage state of this plate due to all these n cracks. If these cracks are far apart, we may say that $B_1 = \bigcup_{i=1}^{n} D_i$, then

$$\mu_{B_1} = \max(\mu_{D_1}, \ldots, \mu_{D_n}) \tag{A-14}$$

If these cracks are fairly close to each other, we may say that $B_2 = \sum_{i=1}^{n} D_i$, then

$$\mu_{B_2} = 1 - \prod_{i=1}^{n}[1 - \mu_{D_i}] \tag{A-15}$$

or,

$$\mu_{B_1} \leqslant \mu_B \leqslant \mu_{B_2} \tag{A-16}$$

More specifically, say that four cracks are detected with the following data: $y_1 = 0.05$, $y_2 = 0.5$, $y_3 = 0.4$, $y_4 = 0.08$. Using Eq. A-13, we find that $\mu_{D_1} = 0$, $\mu_{D_2} = 0.8$, $\mu_{D_3} = 0.6$, then

$$\mu_{B_1} = \max(0, 0.8, 0.6, 0) = 0.8 \tag{A-17}$$

and

$$\mu_{B_2} = 1 - (1 - 0)(1 - 0.8)(1 - 0.6)(1 - 0) = 0.92 \tag{A-18}$$

or,

$$0.8 \leqslant \mu_B \leqslant 0.92 \tag{A-19}$$

In this case, this plate with these four detected cracks is said to have a 'strong' membership in the 'severely damaged' category.

A3 Fuzzy relations

Let P be a product set of n sets and M be its membership set. A fuzzy n-ary relation is a fuzzy set of P taking its values in M. As an example, let $X = \{\text{Building A, Building B}\}$, and $Y = \{\text{Bridge C, Bridge D}\}$. Then, a binary fuzzy relation of 'similar damage state' between members of X and Y may be expressed as

$$R = \begin{array}{c} \\ \text{A} \\ \text{B} \end{array} \begin{array}{c} \begin{array}{cc} \text{C} & \text{D} \end{array} \\ \begin{bmatrix} 0.8 & 0.1 \\ 0.3 & 0.9 \end{bmatrix} \end{array} \tag{A-20}$$

in which the (i,j)th element is the value of the binary membership function $\mu_R(x,y)$ for the ith value of x and the jth value of y. For this numerical example, Building A and Bridge C are said to have a strong membership of 0.8 to be in a similar damage state. These numbers are arbitrarily selected for the purpose of illustration.

The union of two relations, say R and S, is denoted by $R \cup S$ and has the following membership function,

$$\mu_{R \cup S}(x,y) = \mu_R(x,y) \vee \mu_S(x,y) = \max[\mu_R(x,y),\ \mu_S(x,y)] \tag{A-21}$$

where '$\vee$' denotes maximum. On the other hand, the intersection of two relations has the following membership function:

$$\mu_{R \cap S}(x,y) = \mu_R(x,y) \wedge \mu_S(x,y) = \min[\mu_R(x,y),\ \mu_S(x,y)] \tag{A-22}$$

where '$\wedge$' denotes minimum. More generally if R_i, $i = 1, \ldots, n$, are relations, then

$$\mu_{\underset{i}{\cup} R_i}(x,y) = \underset{i}{\vee} [\mu_{R_i}(x,y)] \tag{A-23}$$

$$\mu_{\underset{i}{\cap} R_i}(x,y) = \underset{i}{\wedge} [\mu_{R_i}(x,y)] \tag{A-24}$$

The complement, algebraic sum, algebraic product are represented respectively with the following membership functions:

$$\mu_{\bar{R}}(x,y) = 1 - \mu_R(x,y) \tag{A-25}$$

$$\mu_{R+S}(x,y) = \mu_R(x,y) + \mu_S(x,y) - \mu_R(x,y)\mu_S(x,y) \tag{A-26}$$

$$\mu_{RS}(x,y) = \mu_R(x,y)\mu_S(x,y) \tag{A-27}$$

If R is a fuzzy relation from X to Y, and S is a fuzzy relation from Y to Z, then the composition of R and S is a fuzzy relation which is described with the following membership function:

$$\mu_{R\cdot S}(x,z) = \bigvee_y [\mu_R(x,y) \wedge \mu_S(y,z)] = \max_y [\min(\mu_R(x,y), \mu_S(y,z)] \quad \text{(A-28)}$$

Recall the example relating similar damage states of buildings and bridges as given in Eq. A-20. Let $Z = \{\text{Dam E, Dam F, Dam G}\}$, and

$$S = \begin{array}{c} \\ \text{C} \\ \text{D} \end{array}\begin{array}{c} \begin{array}{ccc} \text{E} & \text{F} & \text{G} \end{array} \\ \begin{bmatrix} 0.7 & 0.5 & 0.4 \\ 0.2 & 0.6 & 0.5 \end{bmatrix} \end{array} \quad \text{(A-29)}$$

Then,

$$R\cdot S = \begin{bmatrix} 0.8 & 0.1 \\ 0.3 & 0.9 \end{bmatrix} \cdot \begin{bmatrix} 0.7 & 0.5 & 0.4 \\ 0.2 & 0.6 & 0.5 \end{bmatrix} \quad \text{(A-30)}$$

$$= \begin{array}{c} \\ \text{A} \\ \text{B} \end{array}\begin{array}{c} \begin{array}{ccc} \text{E} & \text{F} & \text{G} \end{array} \\ \begin{bmatrix} 0.7 & 0.5 & 0.4 \\ 0.3 & 0.6 & 0.5 \end{bmatrix} \end{array}$$

To interpret this result, we say that Building A and Dam E have similar damage state with a degree of membership 0.7, which is obtained from the operation $((0.8 \wedge 0.7) \vee (0.1 \wedge 0.3))$ or $\max(\min(0.8, 0.7), \min(0.1, 0.2))$.

A4 A simple example in structural reliability

For most structures, the calculated probability of failure, p_f, using available statistics is generally smaller than 10^{-6}. Brown (20) indicated that his perceived failure rate is of the order of 10^{-3} for a certain type of structure. Following Blockley (14) in part, Brown (23) applied the theory of fuzzy sets in an attempt to explain the difference between the calculated and observed probabilities of failure. The proposed procedure includes the following steps: (i) compute the objective failure probability $p_F^{(0)} = 10^{-n}$ using all available objective statistical data on load, strength, etc.; (ii) list the gravity g and consequence c, for each subjective factor which can affect the structural safety; (iii) assign linguistic safety statements into fuzzy sets and obtain the total effect $T(g,c)$; (iv) obtain a relation $R(c,n)$ between the consequence of the combined parameters C and safety measure n; (v) find the composite of T and R, i.e., TR, to provide the subjective safety measure for this structure $F_S = F_S(g,n)$; and (vi) extract a subset $K(n)$ from F_S for an element $g \in G$ which yields the fuzzy safety measure.

A simplified version of the numerical examples as given by Brown (20) is presented below for the purpose of illustration. Consider two subjective factors as follows: (1) the effect of mathematical modeling, numerical calculations, and design experience, and (2) the effect of human factors, and construction experience and process. For each factor, the gravity (or importance) of the adverse effect G and the consequence of this effect C are estimated by experts with linguistic statements as listed in Table A1.

Table A1 Estimates for Subjective Factors

Factor	*Gravity, G_i, or importance*	*Consequence, C_i*
(1) Mathematical modeling, numerical calculations, design experience	small	large
(2) Human factors, construction experience and process	medium	grave

To expressly translate linguistic terms such as 'small' in terms of fuzzy sets, let

$$\begin{aligned}
G_1 &= \text{small} = \{(1|0), (0.9|0.1), (0.5|0.2)\} \\
G_2 &= \text{medium} = \{(0.1|0.2), (0.2|0.3), (0.8|0.4), (1|0.5), (0.8|0.6), \\
&\qquad (0.2|0.7), (0.1|0.8)\} \\
C_1 &= \text{large} = \{(0.5|0.8), (0.9|0.9), (1|1)\} \\
C_2 &= \text{grave} = (\text{large})^2 \\
&= \{(0.25|0.8), (0.81|0.9), (1|1)\}
\end{aligned} \tag{A-31}$$

Thus,

$$G_1 \cap C_1 = \begin{array}{c|ccc} & \multicolumn{3}{c}{\text{Consequence = large}} \\ & 0.8 & 0.9 & 1 \\ \hline \text{Gravity = small} \quad 0 & 0.5 & 0.9 & 1 \\ 0.1 & 0.5 & 0.9 & 0.9 \\ 0.2 & 0.5 & 0.5 & 0.5 \end{array} \tag{A-32}$$

$$G_2 \cap C_2 = \begin{array}{c|ccc} & \multicolumn{3}{c}{\text{Consequence = grave}} \\ & 0.8 & 0.9 & 1.0 \\ \hline \text{Gravity = medium} \quad 0.2 & 0.1 & 0.1 & 0.1 \\ 0.3 & 0.2 & 0.2 & 0.2 \\ 0.4 & 0.25 & 0.8 & 0.8 \\ 0.5 & 0.25 & 0.81 & 1 \\ 0.6 & 0.25 & 0.8 & 0.8 \\ 0.7 & 0.2 & 0.2 & 0.2 \\ 0.8 & 0.1 & 0.1 & 0.1 \end{array} \tag{A-33}$$

The total effect of both factors can be obtained by taking the union of Eqs A-32 and A-33, i.e.,

$$T=(G_1\cap C_1)\cup(G_2\cap C_2)= \begin{array}{c|ccc} & \multicolumn{3}{c}{\text{Consequence}} \\ & 0.8 & 0.9 & 1.0 \\ \hline \text{Gravity} & & & \\ 0 & 0.5 & 0.9 & 1 \\ 0.1 & 0.5 & 0.9 & 0.9 \\ 0.2 & 0.5 & 0.5 & 0.5 \\ 0.3 & 0.2 & 0.2 & 0.2 \\ 0.4 & 0.25 & 0.8 & 0.8 \\ 0.5 & 0.25 & 0.81 & 1 \\ 0.6 & 0.25 & 0.8 & 0.8 \\ 0.7 & 0.2 & 0.2 & 0.2 \\ 0.8 & 0.1 & 0.1 & 0.1 \end{array} \qquad \text{(A-34)}$$

To establish a fuzzy relation ($R9c,n$) between the fuzzy sets of consequences C and safety measures N, let

$$R:N=\begin{cases}\text{very large} = (.04\,n),\ (0.64\,n-1),\ (1\,n-2), & \text{if } C \text{ is large}\\ \text{large} = (0.2\,n),\ (0.8\,n-1),\ (1\,n-2), & \text{if } C \text{ is medium}\\ \text{small} = (1\,n),\ (0.8\,n-1),\ (0.2\,n-2), & \text{if } C \text{ is small}\end{cases}$$

Then

$$R_1=C_1\cap N_1= \begin{array}{c|ccc} & \multicolumn{3}{c}{N_1=\text{very large}} \\ & n & n-1 & n-2 \\ \hline \text{Consequence} = \text{large} & & & \\ 0.8 & 0.04 & 0.5 & 0.5 \\ 0.9 & 0.04 & 0.64 & 0.9 \\ 1 & 0.04 & 0.64 & 1 \end{array} \qquad \text{(A-36)}$$

$$R_2=C_2\cap N_2= \begin{array}{c|ccc} & \multicolumn{3}{c}{N_2=\text{large}} \\ & n & n-1 & n-2 \\ \hline \text{Consequence} = \text{medium} & & & \\ 0.2 & 0.1 & 0.1 & 0.1 \\ 0.3 & 0.2 & 0.2 & 0.2 \\ 0.4 & 0.2 & 0.8 & 0.8 \\ 0.5 & 0.2 & 0.8 & 1 \\ 0.6 & 0.2 & 0.8 & 0.8 \\ 0.7 & 0.2 & 0.2 & 0.2 \\ 0.8 & 0.1 & 0.1 & 0.1 \end{array} \qquad \text{(A-37)}$$

$$R_3 = C_3 \cap N_3 = \begin{array}{c|ccc} & & N_3 = \text{small} & \\ & n & n-1 & n-2 \\ \hline \text{Consequence} = \text{small} \quad 0 & 1 & 0.8 & 0.2 \\ 0.1 & 0.9 & 0.8 & 0.2 \\ 0.2 & 0.5 & 0.5 & 0.2 \end{array} \qquad \text{(A-38)}$$

Using Eqs. A-36 through A-38, we obtain

$$R = R_1 \cup R_2 \cup R_3 = \begin{array}{c|ccc} & & N & \\ & n & n-1 & n-2 \\ \hline \text{Consequence} \quad 0 & 1 & 0.8 & 0.2 \\ 0.1 & 0.9 & 0.8 & 0.2 \\ 0.2 & 0.5 & 0.5 & 0.2 \\ 0.3 & 0.2 & 0.2 & 0.2 \\ 0.4 & 0.2 & 0.8 & 0.8 \\ 0.5 & 0.2 & 0.8 & 1 \\ 0.6 & 0.2 & 0.8 & 0.8 \\ 0.7 & 0.2 & 0.2 & 0.2 \\ 0.8 & 0.1 & 0.5 & 0.5 \\ 0.9 & 0.04 & 0.64 & 0.9 \\ 1 & 0.04 & 0.64 & 1 \end{array} \qquad \text{(A-39)}$$

To provide the subjective safety measure, use Eqs. A-34 and A-39 to find the composite as follows:

$$F_S = T \cdot R = \begin{array}{c|ccc} & & N & \\ & n & n-1 & n-2 \\ \hline \text{Gravity} \quad 0 & 0.1 & 0.64 & 1 \\ 0.1 & 0.1 & 0.64 & 0.9 \\ 0.2 & 0.1 & 0.5 & 0.5 \\ 0.3 & 0.1 & 0.2 & 0.2 \\ 0.4 & 0.1 & 0.64 & 0.8 \\ 0.5 & 0.1 & 0.64 & 1 \\ 0.6 & 0.1 & 0.64 & 0.8 \\ 0.7 & 0.1 & 0.2 & 0.2 \\ 0.8 & 0.1 & 0.1 & 0.1 \end{array} \qquad \text{(A-40)}$$

From Eq. 40, choose a subset $K(n)$ of the composition F_S, where $K(n)$ is called a fuzzifier. As an example, choose the largest element in each column and we have,

$$F = \{(n|0.1), (n-1|0.64), (n-2|1)\} \tag{A-41}$$

In this case, if the objective failure probability is 10^{-6}, the inclusion of subjective factors produces a failure probability of the order of 10^{-4} which is closer to Brown's perceived value.

In summary, two subjective factors are evaluated linguistically in terms of the gravity and consequency of each factor as shown in Table A1. Appropriate membership functions are then assigned to such linguistic descriptions in Eq. A-31. The total effect of both factors is given in Eq. A-34. A fuzzy relation between the consequence and a safety measure is established in Eq. A-39. Finally, the fuzzy relation between the gravity and the safety measure is obtained by taking composition of Eqs. A-34 and A-39. Results of this example are used to illustrate a rational evaluation of subjective factors in the structural reliability analysis.

A5 Additional examples

There are two types of data from the inspection and testing of the structure. One type of observation is made from local phenomena such as cracks in certain structural members. Such information can be incorporated in a logical manner to obtain an estimate of the damage state of the whole structure. The other type of data are taken from global behavior of the structure such as the structural response and ground-motion records.

Let B denote the event that the whole structure has been severely damaged, and B_i denote the severely-damaged state of the structure using the ith group of data. For example, $i = 1$ corresponds to the information on detected cracks in the structure, and $i = 2$ corresponds to the features extracted from recorded accelerograms. Therefore, for m groups of data, we have,

$$B = \bigcup_{i=1}^{m} B_i \tag{A-42}$$

or,

$$\mu_B = \bigvee_i (\mu_{B_i}) \tag{A-43}$$

Furthermore, for the ith group of data which are related to the jth component of the structure consisting of a total of n components, let D_{ij} denote the severely damaged state of the jth component. Then B_i can be considered as the algebraic sum of the damage of each component, i.e.,

$$B_i = \sum_{j=1}^{n} D_{ij} \tag{A-44}$$

or

$$\mu_{B_i} = 1 - \prod_{j=1}^{n} [1 - \mu D_{ij}] \tag{A-45}$$

For the purpose of illustration as noted above, let B_1 denote the severely-damaged state of the structure from crack detection and measurements, and B_2 denote the severely-damaged state of the structure from a reduction of the natural (fundamental) frequency of the structure. Say that there are three major components with detected cracks, and we have $\mu_{D_{11}} = 0$, $\mu_{D_{12}} = 0.8$, $\mu_{D_{13}} = 0.6$, then

$$\mu_{B_1} = 0.92 \tag{A-46}$$

Meanwhile, we find that the calculated reduction of measured natural frequency is 25%. Through the use of an hypothetically established membership function, we obtain

$$\mu_{B_2} = 0.78 \tag{A-47}$$

The determination of this membership can be based on full-scale destructive test data such as those of Galambos and Mayes (61) plus advice from various experts. Then, the membership of the structure in the severely-damaged state is given by

$$\mu_B = \max(\mu_{B_1}, \mu_{B_2}) = 0.92 \tag{A-48}$$

As another possible approach, let $X = \{x_1, x_2, \ldots, x_k\}$ be a set of k features. For example, x_1 = many cracks, x_2 = large cracks, and x_3 = excessive deformation. Also, let $Y = \{y_1, y_2, \ldots, y_l\}$ be a set of l potential failure modes. For example, y_1 = fatigue and fracture failure, y_2 = creep, y_3 = instability, and y_l = progressive collapse. Furthermore, let Z = the severely-damaged state. If we can find the fuzzy relations R (from X to Y) and S (from Y to Z), we can relate features X to the severely-damaged state of the structure Z by taking the composition RS. For the purpose of illustration, let R and S be given as follows:

	y_1: Fatigue and fracture	y_2: Creep	y_3: Instability	y_4: Progressive collapse
x_1: many cracks	0.9	0.2	0.4	0.4
$R =$ x_2: large cracks	0.8	0.3	0.7	0.8
x_3: excessive deformation	0.3	0.8	0.9	0.7

(A-49)

$$S = \begin{array}{c|c} & z \\ & \text{severely damaged} \\ \hline y_1 & \\ y_2 & \\ y_3 & \\ y_4 & \end{array} \begin{bmatrix} 0.4 \\ 0.3 \\ 0.8 \\ 1.0 \end{bmatrix} \tag{A-50}$$

Then

$$R \cdot S = \begin{array}{c|c} & z \\ \hline x_1 & \\ x_2 & \\ x_3 & \end{array} \begin{bmatrix} 0.4 \\ 0.8 \\ 0.8 \end{bmatrix} \tag{A-51}$$

Results as given in Eq. A-51 indicate that the presence of features x_2 (large cracks) and x_3 (excessive deformation) would constitute a strong membership of the structure being in the severely damaged state. In other words, if large cracks and excessive deformations are present, the structure can be classified as being 'severely damaged'.

Appendix B **SPERIL-I**

B1 General remarks

The name of this program is taken from *S*tructural *PERIL*. Professor Mitsuru Ishizuka of the University of Tokyo visited Purdue University from 1980 to 1981, where he collaborated with Professors K. S. Fu (56–58) and J. T. P. Yao to produce this preliminary version I (80, 81).

SPERIL is a computer-based damage assessment system of existing structures, which are subjected to strong earthquake excitations. To classify the damage state of a given structure, SPERIL can be used to interpret observed data including analytical results of accelerometer records and on-site visual inspection.

The expert system approach, with which a complex decision-making problem can be decomposed into several simpler hierarchial subproblems and useful knowledge can be collected as fragmentary rules, has been used to utilize experienced engineers' knowledge. This approach can be used to provide the capability of dealing with a great variety of structural conditions in the structural damage assessment problem. Because the related knowledge often contains uncertainty and impreciseness of expression range, a new inference procedure with uncertainty and fuzzy restriction has been developed. The inference proceeds to obtain certainty measures at higher subgoals by using available rules and observed data, and eventually gives certainty measures at final goal state, which suggest an appropriate answer. In SPERIL version I, separate evidential observations are integrated on the basis of Dempster & Shafer's theory for fuzzy subsets.

It is noted that the SPERIL program (version I) is a preliminary version. The rules as written in rule-base are being refined and updated with more accurate and specific rules in current investigations. Nevertheless this first version does demonstrate the feasibility of a systematic computer-based damage assessment system. The program portion is written by using language C with some 8000 statements. On the other hand, the rules in the rule-base are written in a format close to natural rule sentences.

B2 Rule representation

The whole program configuration of SPERIL consists of three separate files. Rule-base (file name: Rbase) is a completely separate storage from control and inference process. Useful knowledge for the inference purpose has been collected and expressed in a stylized rule format. The rule format is so designed that both human and computer can interpret it easily. First line of the rule is Rule and followed with a 4-digit rule number, the first two digits of which are rule set number corresponding to the node numbers as shown in Fig. B1. Following the first line, rule statements are written in as many lines as is necessary. Each statement line has headers such as field-1, field-2 and field-3. Eleven types of headers as listed in Table B1 are recognized by the computer. The end of one rule is found by encountering an unrecognizable header. The line with header type 18 is premise statement; and the line with header type 9 and 10 is action statement. The fundamental function of production system, that is, 'if premise is satisfied then action is taken' is emphasized in rule interpretation. The action in this case is an updating process of short term memory.

Table B1 Defined Header of Rule Statement

Header	*Header type*
IF	1
OR IF	2
THEN IF	3
ELSE IF	4
THEN	9
ELSE	10
and	5 (in the context of 1.2)
	7 (in the context of 3.4)
	11 (in the context of 9)
or	6 (in the context of 1.2)
	8 (in the context of 3.4)

In the field-1, the name of short term memory is written. The interpretation of field-2 and field-3 is determined according to the type of header and type of short term memory which will be described in detail in the following section. In the present version of SPERIL, the combinations of Table B2 are defined and implemented for the processing. The interpretations of these statements are straightforward. As fuzzy subsets which represent the fuzzy grade of damage in this particular case, acceptable expressions are restricted to six, namely, no, slig (slight), mode (moderate), seve (severe), dest (destructive) and uk (unknown). If the statement

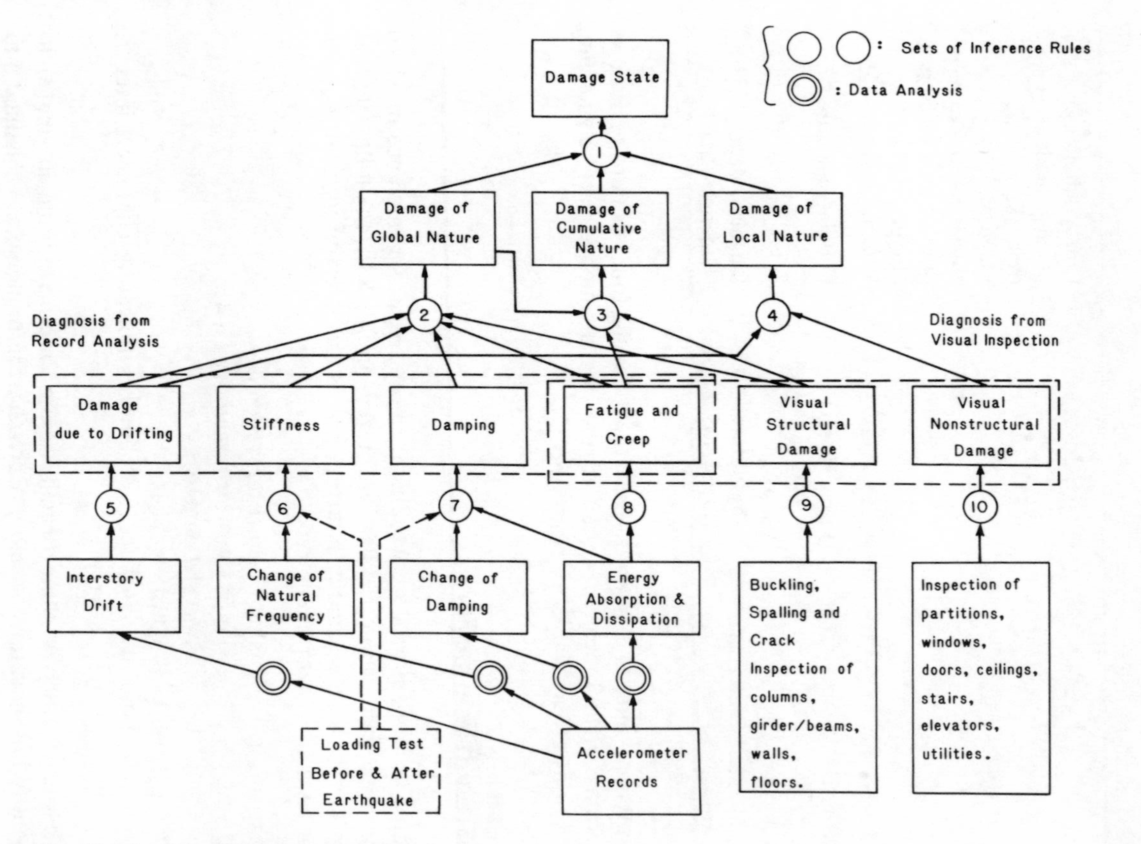

Fig. B1 Inference network of SPERIL

Table B2 Defined Rule Statement in SPERIL

Header	*Type of short term memory (field-1)*	*Field-2*	*Field-3*
Premise	1	is	(Fuzzy subset) no, slig, mode, seve, dest, uk
	2	is	Linguistic data
	3	>=, >, <=, <	Numerical data or name of short term memory of type-3
	4	is	Yes, no
Action	1	(Fuzzy subset) no, slig, mode, seve, dest, uk	Certainty measure
	2	= (Substitute)	Linguistic data
	3	=, ++ (Accumulate)	Numerical data

is premise and has 't' in position 21, the following four positions perform as optional field-3, which is used to indicate a union subset. Remaining positions are available for comments.

B3 Short term memory

Short term memories are working memory spaces for inference in which the input data or inferred data are stored. In SPERIL, the following four (in fact 13 because smd is an array) spaces are reserved in main program for each short term memory:

char smn (3) :name of short term memory
int smt :type of short term memory (1–4)
float smd (10) :numerical data array
char sml (4) :linguistic data
(char, int. float are language C definition indicating character, integer, floating point, respectively)

Three characters name (usually upper case) is written in smn(3) to identify the short term memory. Each short term memory is classified into four types (1–4), one of which is indicated in each smt. Table B3 shows those types of short term memory and the meanings of their memory contents. Each character memory sml is initialized to 'unan' to serve as an indicator showing that the short term memory has not been written as yet.

Table B3 Types of Short Term Memories

Type	*Meaning of memory contents*
1	Certainty measures of fuzzy subsets:
	no : smd[0] mode seve : smd[5]
	no slig : smd[1] seve : smd[6]
	slig : smd[2] seve dest : smd[7]
	slig mode : smd[3] dest : smd[8]
	mode : smd[4] uk : smd[9]
2	Linguistic data (up to 4 characters) : sml
3	Numerical data : smd[0]
4	Yes—no data
	yes : smd[0] = 1
	grey : smd[0] = 0.6, smd[1] = 0.2, smd[2] – 0.2
	no : smd[1] = 1
	uk : smd[2] = 1

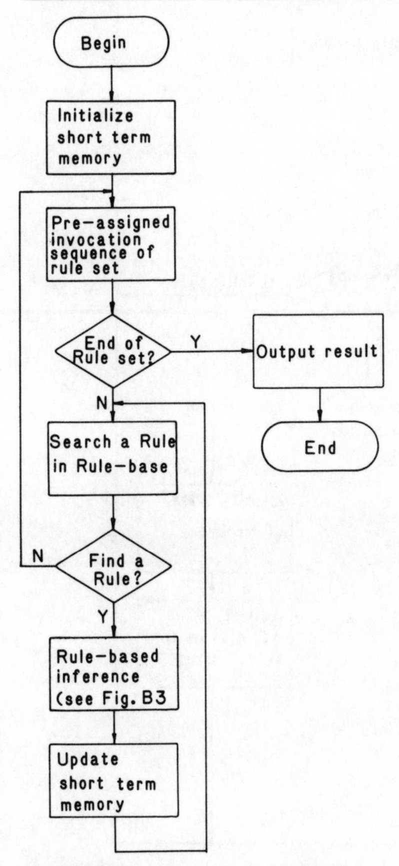

Fig. B2 Main control flow of SPERIL

Whenever the premise is examined or action is taken, the type of short term memory is referred to proceed to an appropriate interpretation of the rule statement.

B4 Control and inference

Figures B2 and B3 show the control and rule-based inference flow of SPERIL. Because the inference network is not deep, no heuristic or

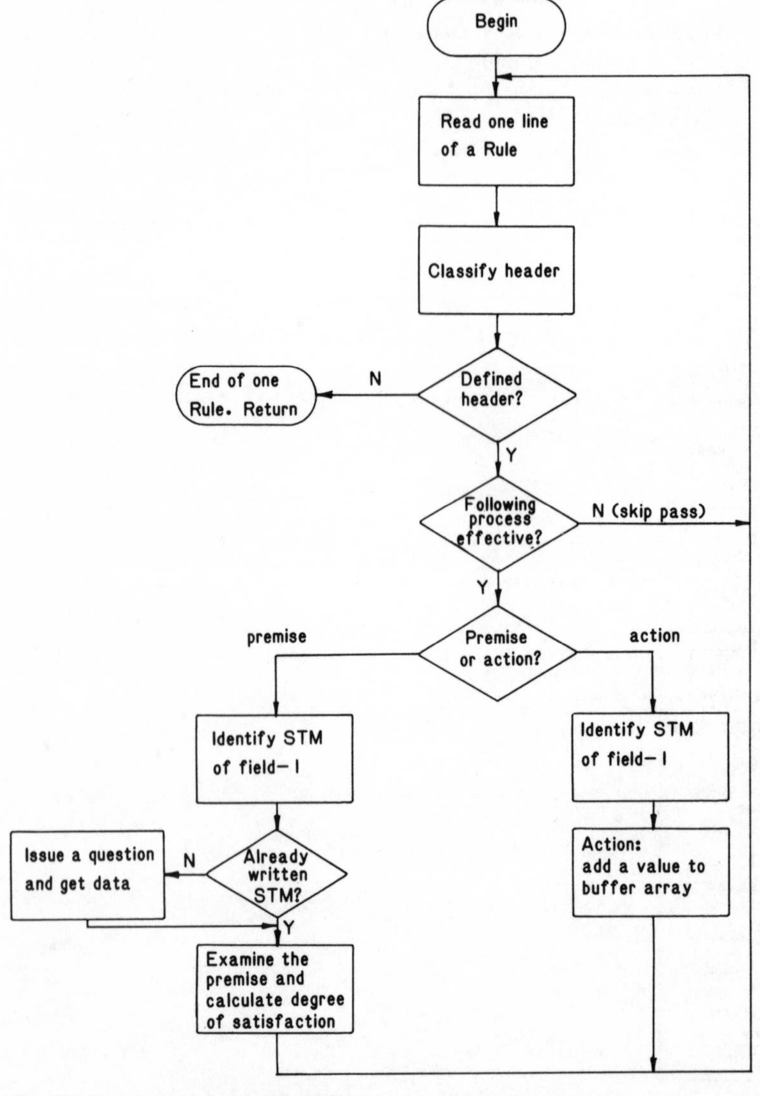

Fig. B3 Processing of one rule (STM: short term memory)

sophisticated strategy of rule invocation is adopted. The sequence of rule set invocation is pre-assigned as follows:

'05', '06', '07', '08', '09', '10', '02', '03', '04', '01'.

This corresponds to bottom-up search rather than top-down or goal-oriented research.

If a relating rule is found in the rule-base, it is processed according to Fig. B2. When the premise is examined and the associated short term memory is found to be unwritten or unanswered, a question is issued to get data. The question is generated referring to a question file (file name: quest) in which an appropriate question sentence is stored for each short term memory with possibility of getting data from operator rather than the inference process.

To avoid the issue of annoying unnecessary questions, skip pass is provided for the case that there is no possibility for later action statements to be taken. Thus only minimum necessary questions are issued for the purpose of inference.

After one rule is processed, the resultant content of buffer memory is used to update the short term memory indicated by field-1 of the action statement. For type-1 short term memories, the Dempster & Shafer's theory extended to fuzzy subset is employed for their updating or combining separate evidences.

Final decision is made according to Dempster & Shafer's lower probabilities of the fuzzy subsets in final goal (FIN) which is damage state. If no fuzzy subset has larger lower probability than a certain threshold (0.2), SPERIL selects no appropriate answer. Therefore, the answer is one of the following:

(1) no damage,
(2) slight damage,
(3) moderate damage,
(4) severe damage,
(5) destructive damage,
(6) no appropriate answer.

B5 Source program

(Demonstration purposes only)

```
/****** speril.c    ----    SPERIL (version I) main program ******/
/*      with files of rbase(rule-base) and quest(questions)      */

#include <stdio.h>

  /* short term memory */
     /* smt[i]           */
     /*  1       smd[0-9]:certainty measures (Dempster & Shafer's
                          basic prob.) of no, no-slig, slight,
                          slig-mode, moderate, mode-seve, severe,
                          seve-dest, destructive, unknown.
         2        *sml   :linguistic data.
         3        smd[0] :numerical data.
         4       smd[0-2]:yes, no, unknown.   */
#define NSTM 50
  char  smn[NSTM][4]  ;        /*name*/
  int   smt[NSTM];            /*type*/
  float smd[NSTM][10];        /*numerical data*/
  char  sml[NSTM][5]  ;        /*linguistic data*/
  int   pp;                   /*control of print-out*/

main()
{
  char bname[50], date[10];
  char *seq[15];
  char rsno[3];  /*rule set no.*/
  int i=0;
  extern int pp;

  /*sequence of rule set invocation*/
  seq[0]="05";
  seq[1]="06";
  seq[2]="07";
  seq[3]="09";
  seq[4]="10";
  seq[5]="02";
  seq[6]="04";
  seq[7]="01";
  seq[8]="$$";

  printf("****************** SPERIL  (Version I) ******************** 0);
  printf(" Rule-based Damage Assessment System for Existing Structure 0);
  printf(" subjected to Earthquake Excitation 0);
  printf("***********************************************************0);
  printf("Answer the following questions.0);
  printf("Then I can suggest a reasonable answer. 0);

  printf("Enter the name of building. =");
  scanf("%s",bname);
  printf("Enter date. = ");
  scanf("%s",date);
  printf("Do you want to see rule sentences on display {y or n} ? =");
  scanf("%s",rsno);  pp=rsno[0];
```

```
  initial();                              /*initialize*/

  /*main loop*/
  while (*seq[i] != '$')
   { rsno[0] = *seq[i];
     rsno[1] = *(seq[i]+1);
     rsno[2] = *(seq[1]+2);                /* ' ' : end of string*/

     rbinfer(rsno);                      /****rule-based inference****/
     i++;
   }

  result();
}




/*** initialize short term memory ***/
initial()
{
  extern char  *smn[];
  extern int    smt[];
  extern float  smd[][10];
  extern char  *sml[];
  int i, n, j;

  smn[0]="FIN";  smt[0]=1;  sml[0]="unan";
  smn[1]="GLO";  smt[1]=1;  sml[1]="    ";
  smn[2]="CUM";  smt[2]=1;  sml[2]="    ";
  smn[3]="LOC";  smt[3]=1;  sml[3]="    ";
  smn[4]="STI";  smt[4]=1;  sml[4]="    ";
  smn[5]="DAM";  smt[5]=1;  sml[5]="    ";
  smn[6]="FAT";  smt[6]=1;  sml[6]="    ";
  smn[7]="VST";  smt[7]=1;  sml[7]="    ";
  smn[8]="VNS";  smt[8]=1;  sml[8]="    ";
  smn[9]="MAT";  smt[9]=2;  sml[9]="    ";
  smn[10]="INS";  smt[10]=2;  sml[10]="    ";
  smn[11]="FLX";  smt[11]=2;  sml[11]="    ";
  smn[12]="N01";  smt[12]=2;  sml[12]="    ";
  smn[13]="N02";  smt[13]=2;  sml[13]="    ";
  smn[14]="N03";  smt[14]=2;  sml[14]="    ";
  smn[15]="N04";  smt[15]=2;  sml[15]="    ";
  smn[16]="N05";  smt[16]=2;  sml[16]="    ";
  smn[17]="N06";  smt[17]=2;  sml[17]="    ";
  smn[18]="N07";  smt[18]=2;  sml[18]="    ";
  smn[19]="N08";  smt[19]=2;  sml[19]="    ";
  smn[20]="N09";  smt[20]=2;  sml[20]="    ";
  smn[21]="N10";  smt[21]=2;  sml[21]="    ";
  smn[22]="N11";  smt[22]=2;  sml[22]="    ";
  smn[23]="ISD";  smt[23]=3;  sml[23]="    ";
  smn[24]="NST";  smt[24]=3;  sml[24]="    ";
```

```
  smn[25]="CNF";  smt[25]=3;  sml[25]="    ";
  smn[26]="---";               /*killed*/
  smn[27]="CDI";  smt[27]=3;  sml[27]="    ";
  smn[28]="CDD";  smt[28]=3;  sml[28]="    ";
  smn[29]="SUM";  smt[29]=3;  sml[29]="    ";
  smn[30]="S01";  smt[30]=4;  sml[30]="    ";
  smn[31]="S02";  smt[31]=4;  sml[31]="    ";
  smn[32]="S03";  smt[32]=4;  sml[32]="    ";
  smn[33]="S04";  smt[33]=4;  sml[33]="    ";
  smn[34]="S05";  smt[34]=4;  sml[34]="    ";
  smn[35]="S06";  smt[35]=4;  sml[35]="    ";
  smn[36]="S07";  smt[36]=4;  sml[36]="    ";
  smn[37]="C01";  smt[37]=4;  sml[37]="    ";
  smn[38]="C02";  smt[38]=4;  sml[38]="    ";
  smn[39]="C03";  smt[39]=4;  sml[39]="    ";
  smn[40]="C04";  smt[40]=4;  sml[40]="    ";
  smn[41]="C05";  smt[41]=4;  sml[41]="    ";
  smn[42]="C06";  smt[42]=4;  sml[42]="    ";
  smn[43]="INF";  smt[43]=3;  sml[43]="    ";
  smn[44]="IFS";  smt[44]=3;  sml[44]="    ";
  smn[45]="DRI";  smt[45]=1;  sml[45]="    ";
  smn[46]="S08";  smt[46]=4;  sml[46]="    ";
  smn[47]="$$$";

  for(i=0;i < NSTM ;i++)
    { for(j=0;j<=4;j++) *(sml[i]+j) = *(sml[0]+j);
      for(j=0;j<=9;j++) smd[i][j]=0.0; }
}


/********** rule-based inference **********/
rbinfer(rsno)
   char rsno[];
{
  FILE *fprb;
  extern char  *smn[];
  extern int   smt[];
  extern float smd[][10];
  extern char  *sml[];
  extern int   pp;
  char line[100];
  char buf3[3], buf4[4];
  int  hty, preh, nact, im;
  int  i, jj, k,  itype , skip;
  float dd, dsatis, dfif, dsif, ssatis, sact=0;
  float premise(), min(), max();
  float bp[6];       /*basic pro. of no,slig,mode,seve,dest,uk*/
```

```
if(pp=='t') printf("     Rule Set # = %s 0,rsno);

fprb=fopen("rbase","r");

do                        /*loop until break*/
{
                          /*find associated rule in rule-base*/
   k=0;
   while (k == 0)
      k=findr(fprb,rsno);    /*k=1 if rule is found*/
   if(k <= -1) break;       /*k=-9 if Rule$$.. is detected*/

   for (i=0; i<=9; i++) bp[i]=0;
   nact=0;  preh=1;  dfif=1;  skip=0;

   do                        /*loop until break*/
    {
     lget(fprb,line);        /* 1 line get*/

     hty=header(line,preh); /*header classify*/
     preh=hty;
     if(hty <= 0) break;     /*undefined header*/

     if(hty==3) dfif=dsatis;           /*THEN IF  set first IF*/
     if(hty==3 && dfif<0.1) skip=1;  /*THEN IF &&small first IF*/

     if((skip==1 && hty != 2 && hty!=10) || ssatis>0.99)
         { dsatis=0;                   /*skip examination*/
           goto skip; }

     if(pp=='y' || pp=='t') printf("     %s", line);

     if (hty==1 || hty==2)
         {dsatis=1; ssatis=0;      /*reset for IF or OR IF*/
          skip=0; }                /*no OR IF in present rule base*/

     if (hty <= 8)            /*premise clause*/
        dd=premise(line);     /*deg. of satisfaction*/

     if(hty==1 || hty==2 || hty==5)     /*IF, IF-and*/
           dsatis=min(dsatis,dd);
     else if(hty==6)                    /*IF-or*/
           dsatis=max(dsatis,dd);

     else if(hty==3 || hty==4)          /*THEN IF or ELSE IF*/
           {dsif=dd;          /*deg. of second IF clouses*/
            dd=min(dfif,dsif);
            dsatis=min(dd,(1-ssatis)); }
     else if(hty==7)               /*ELSE IF-and*/
           {dsif=min(dsif,dd);
            dd=min(dfif,dsif);
            dsatis=min(dd,(1-ssatis)); }
     else if(hty==8)               /*ELSE IF-or*/
```

```
              {dsif=max(dsif,dd);
               dd=min(dfif,dsif);
               dsatis=min(dd,(1-ssatis)); }
if(pp=='t' && hty<=8)
printf("func.rbinfer:IF: hty=%d dd=%5.2f ssatis=%5.2f dsatis=%5.2f0
  ,hty,dd,ssatis,dsatis);    /***test***/

     skip:   ;

     if(hty==9 && nact==0)    /*first action - identyfy stm*/
        { nact++;
          stmove(line,8,3,buf3);
          im=findstm(buf3);
          itype=smt[im];  }

     if(hty==9 && (itype==1 || itype==3))         /*THEN*/
          {
           action(line,dsatis,&sact,bp);    /*action*/

           ssatis=ssatis+dsatis;
if(pp=='t')
printf("func.rbinfer:THEN: dsatis=%5.2f ssatis=%5.2f sact=%5.2f 0
  ,dsatis, ssatis, sact); /***test***/
           nact++;
          }

     else if(hty==11 && itype==1)              /*THEN-and*/
          {action(line,dsatis,&sact,bp);
           nact++; }

     else if(hty==9 && itype==2)        /*substitute linguistic data*/
          { if(dsatis > 0.99)
              { stmove(line,12,3,buf3);
                if(comp(buf3,"=  ",3)==1)
                stmove(line,17,4,sml[im]);
                ssatis=ssatis+dsatis,  nact++;
                sact=sact+dsatis;
    if(pp=='t') psml(); /***test***/
              } }
     else if(hty==10)                   /*ELSE*/
           bp[5]=1-sact;                   /*unknown*/

     }
   while(hty > 0);

/*end of one rule.  update short term memory*/

if(pp=='t' && dsatis>0.1)
printf("func.rbinfer: ssatis=%f sact=%f0, ssatis,sact); /*test***/

   if (itype==1 && ssatis>=0.1)        /*neglect small contribution*/
          update(im,bp);       /*Dempster's rule of combination*/

    else if(itype==3 && ssatis>0.1)             /*for numerical data*/
      { smd[im][0]=smd[im][0]+bp[0];
        sml[im]="answ"; }
```

```
      if((pp=='y' || pp=='t') && ssatis>0.1)
        { printf("     %s : ", smn[im]);
          if(itype==1)
            { for(i=0;i<=5;i++) printf("%5.2f      ",bp[i]);
              printf("                ");
              for(i=0;i<=9;i++) printf("%5.2f", smd[im][i]); }
          if(itype==3)
            { printf("%8.3f0,bp[0]);
              printf("                %8.3f", smd[im][0]); }
          if(itype==2)  printf("%s",sml[im]);
          printf("0); }

  }
  while( k >= 0);

  fclose(fprb);
}



/******** check premise ******** --- return deg. of satisfaction */
float premise(line)
  char line[];
{
  extern int  smt[];
  extern char *sml[];
  extern int  pp;
  char buf3[3];
  int im, itype, unans;
  float degs, deg1(), deg2(), deg3(), deg4();

  stmove(line,8,3,buf3);
  im=findstm(buf3);      /*find stm no.*/
  itype=smt[im];         /*read stm type (1,2,3,4)*/

  unans=ccomp( sml[im], "unan", 4);
                         /* 1 if unanswered */
  if(pp=='t') printf("func.premise:sml[%d]=%s$ unans=%d0,im,sml[im],unans);
  if(unans==1)  ask(im,itype);  /*issue a question*/

  if(itype==1)      degs=deg1(im,line);
  else if(itype==2) degs=deg2(im,line);
  else if(itype==3) degs=deg3(im,line);
  else if(itype==4) degs=deg4(im,line);

  if(pp=='t')
  printf("func.premise: im=%d  itype=%d  degs=%7.3f 0, im, itype, degs); /*
  return(degs);
}
```

```
/********** degree of satisfaction for type-2 **********/
float deg2(im,line)
  int im;
  char line[];
{
  extern char *sml[];
  char buf4[4];
  int i, j, jj;
  float degs;

  stmove(line,17,4,buf4);
  for(i=0;i<=3;i++)
    { if(buf4[i]=='0 || buf4[i]=='0')
        for(j=i;j<=3;j++) buf4[j]=' '; }
  /*linguistic data*/
  if((jj=comp(buf4, sml[im], 4)) == 1) degs=1;
  else degs=0;
  return(degs);
}


/********** degree of satisfaction for type-3 **********/
float deg3(im,line)
  int  im;
  char line[];
{
  extern int   smt[];
  extern float smd[][10];
  extern char  *sml[];
  char buf6[6];
  int jj, imm, itype;
  float xdata, x, degs=0, convert();

  stmove(line,17,6,buf6);

  if ((buf6[0] >= '0' && buf6[0] <= '9') || buf6[0] == '-' )
                                         /*numerical data*/
      x=convert(buf6);                   /*convert char to float*/

  else                        /*character -- search short term memory*/
    { imm=findstm(buf6);
      itype=smt[imm];
      if (ccomp(sml[imm], "unan",4) == 1)
         ask(imm,itype);
      x=smd[imm][0]; }

  stmove(line,12,2,buf6);
  xdata=smd[im][0]; /*smd[im][0] is input numerical data*/
```

```
  if (ccomp(buf6,">=",2) == 1)
    { if (xdata >= x) degs=1; }
  else if (ccomp(buf6,"> ",2) == 1)
    { if (xdata > x) degs=1; }
  else if (ccomp(buf6,"<=",2) == 1)
    { if (xdata <= x) degs=1; }
  else if (ccomp(buf6,"< ",2) == 1)
    { if (xdata < x) degs=1; }

  return(degs);            /*degs=0.0 or 1.0*/
}


/********** degree of satisfaction for type-4 **********/
float deg4(im,line)
  int im;
  char line[];
{
  extern float smd[][10];
  char buf4[4];
  int jj;
  float degs;

  stmove(line,17,4,buf4);

  if ((jj=comp(buf4,"yes",3)) == 1)  degs=smd[im][0];
  else if ((jj=comp(buf4,"no",2)) == 1) degs=smd[im][1];
  else  degs=0;

  return(degs);
}


/********** degree of satisfaction for type-1 **********/
/*   certaimty measures of no, no-slig, slig,..... */
float deg1(im,line)
  int im;
  char line[];
{
  extern float smd[][10];
  char buf4[4];
  int i;
  float degs=0, bp[10], flow();

  for(i=0; i<=9; i++) bp[i]=smd[im][i];
  stmove(line,17,4,buf4);
  degs=flow(buf4,bp);          /*lower prob. of fuzzy subset*/

  if (line[21] == '+')
      {stmove(line,22,4,buf4);
       degs=degs+flow(buf4,bp); }

  return(degs);

  }
  /*************** lower prob. of fuzzy subset****************/
  float flow(buf,bp)
    char buf[];
```

```
  float bp[];
{
  float lp;

  if (ccomp(buf,"no",2) == 1)      lp=bp[0]+bp[1];
  else if (ccomp(buf,"slig",2) == 1) lp=bp[1]+bp[2]+bp[3];
  else if (ccomp(buf,"mode",2) == 1) lp=bp[3]+bp[4]+bp[5];
  else if (ccomp(buf,"seve",2) == 1) lp=bp[5]+bp[6]+bp[7];
  else if (ccomp(buf,"dest",2) == 1) lp=bp[7]+bp[8];
  else  lp=0;

  return(lp);
}

/********** lower prob. *******************/
float lp(im,i)
  int im, i;
{
  extern float smd[][10];
  float llp;
  if(i==0) llp=smd[im][0]+smd[im][1];
  if(i==1) llp=smd[im][1]+smd[im][2]+smd[im][3];
  if(i==2) llp=smd[im][3]+smd[im][4]+smd[im][5];
  if(i==3) llp=smd[im][5]+smd[im][6]+smd[im][7];
  if(i==4) llp=smd[im][7]+smd[im][8];
  if(i==5) llp=smd[im][9];                  /*uk*/
  return(llp);
}


/********** issue a question and get data **********/
ask(im,itype)
  int im, itype;
{
  extern char  *smn[];
  extern float  smd[][10];
  extern char  *sml[];
  extern int pp;
  FILE *fpq;
  char line[100], buf4[4];
  int i, j, jj, jk;
  float dd;

  if ((itype<=1) || (itype>=5)) return;

  fpq=fopen("quest","r");       /*stored file is "quest"*/

  jk=0;
```

```
  while(jk==0)                          /*search desired question*/
   {
    lget(fpq,line);
    if (ccomp(line,"Q-",2) == 1)
        {stmove(line,2,3,buf4);
         if (ccomp(buf4, smn[im], 3) == 1) jk=1;    /*find*/
         else if(ccomp(buf4,"$$$",3) == 1) jk = -1;     /*no quest*/
        }
    }
  if (jk == -1)
      {printf("no question setence in file-quest0);
       return; }

  lget(fpq,line);
  do { printf("%s", line);          /*print question*/
       lget(fpq,line); }
  while (ccomp(line,"Q-",2) != 1);

  fclose(fpq);
  printf("===");
  stmove("an##",0,4,sml[im]);

 /* * * data input * * */
  if(itype==2)                          /*linguistic data*/
    {scanf("%s",line);
     for(i=0;i<=3;i++)
        { if(line[i]=='0 || line[i]==' ')
             for(j=i;j<=3;j++) line[j]=' ';
        }
     stmove(line,0,4,sml[im]);
    }
  else if(itype==3)                     /*numerical data*/
    {scanf("%f", &dd);
     smd[im][0]=dd;
    }
  else if(itype==4)                     /*yes, grey, no, uk*/
    {scanf("%s",line);
     if ((jj=comp(line,"yes",1)) == 1)
         smd[im][0]=1;
     if ((jj=comp(line,"grey",1)) ==1)
         {smd[im][0]=0.6;
          smd[im][1]=0.2; }             /*sum is not 1.0*/
     if ((jj=comp(line,"no",1)) == 1)
          smd[im][1]=1;
     if ((jj=comp(line,"unknown",1)) ==1)
          smd[im][2]=1;
    }
}



/********** action THEN clause **********/
```

```
action(line, dsatis, sact, bp)
  char line[];
  float dsatis, *sact, bp[];
{
  char buf4[4], buf6[6];
  float cm, convert();

  if (dsatis < 0.1) return;          /*neglect small contribution*/

  stmove(line,17,6,buf6);
  cm=convert(buf6);                  /*convert written cm to float*/
  cm=cm*dsatis;

  if (cm < 0.1) return;              /*neglect small contribution*/

  *sact = *sact + cm;
  stmove(line,12,4,buf4);

  if      (ccomp(buf4,"no  ",2) == 1)
      bp[0]=bp[0]+cm;
  else if(ccomp(buf4,"slig",2) == 1)
      bp[1]=bp[1]+cm;
  else if(ccomp(buf4,"mode",2) == 1)
      bp[2]=bp[2]+cm;
  else if(ccomp(buf4,"seve",2) == 1)
      bp[3]=bp[3]+cm;
  else if(ccomp(buf4,"dest",2) == 1)
      bp[4]=bp[4]+cm;
  else if(ccomp(buf4,"uk  ",2) == 1)
      bp[5]=bp[5]+cm;

  else if(ccomp(buf4,"=   ",2) == 1)     /*not certainty measure*/
      bp[0] = cm ;                  /*substitute numerical to smd[im][0]*/
  else if(ccomp(buf4,"++  ",2) == 1)
      bp[0] = bp[0] + cm;           /*accumulate*/
}


/********** update short term memory **********/
/*   Dempster's rule of combination for fuzzy subset */
update(im, bp)
  int im;
  float bp[];
{
  extern float smd[][10];
  extern char  *sml[];
  int i, ii, jj;
  float dbp[10], pp[10], sum;

  if(bp[5] > 0.95) return;                  /*almost unknown*/

  sum=0;

  for(i=0;i<=5;i++) sum=sum+bp[i];
  if(sum<0.999) bp[5]=bp[5]+(1-sum);        /*increase unknown*/

  if(ccomp(sml[im],"unan",4) == 1)  /*first update*/
```

```
    {for(i=0;i<=5;i++)                              /*simple insert*/
       { ii = 2*i;
         if(ii == 10) ii=9;
         smd[im][ii]=bp[i];
       }
     stmove("an$$",0,4,sml[im]);
    }

  else            /*Dempster's rule of combination for fuzzy subset*/
    {
     for(i=0;i<=9;i++) dbp[i]=smd[im][i];

     for(i=0;i<=4;i++)
       {ii=2*i;
        pp[ii]=dbp[ii]*bp[i]+dbp[9]*bp[i]+dbp[ii]*bp[5];
       }
     pp[9]=dbp[9]*bp[5];                             /*unknown set*/

     for(i=0;i<=4;i++)                               /*intersect fuzzy set*/
       {ii=2*i-1;
        pp[ii]=0.5*(dbp[ii-1]*bp[i+1]+dbp[ii+1]*bp[i]);
       }

     sum=0;                 /*normalize*/
     for(i=0;i<=8;i++) sum=sum+pp[i];
     if(sum < 0.05)         /*no normalize if small certainty*/
         pp[9]=1.0-sum;
     else
        {sum=sum+pp[9];
         for(i=0;i<=9;i++) pp[i]=pp[i]/sum;
        }

     for(i=0;i<=9;i++)      /*write to short term memory*/
        smd[im][i]=pp[i];
    }
}


/********** report the result **********/
result()
{
  float  lp(),  lpmax=0;
  int  i, j, im;

  im=findstm("FIN");

  for(i=0;i<=4;i++)              /*search max*/
     { if( lp(im,i) > lpmax)
        { lpmax=lp(im,i);
           j=i; }
     }
```

```
  printf("0*****************************************************0);

  if(lpmax < 0.2)
    printf("  An appropriate answer is not obtained.0);

  else
    {if(j == 0) printf("  There is no damage.0);
     if(j == 1) printf("  The damage is slight.0);
     if(j == 2) printf("  The damage is moderate.0);
     if(j == 3) printf("  The damage is severe.0);
     if(j == 4) printf("  The damage is destructive.0);
    }

  printf("0*****************************************************0);

  printf("          no   slight moderate   severe destruct   unknown 0);
  printf("FIN");
  for(i=0;i<=5;i++) printf("%9.4f",lp(im,i));
  im=findstm("GLO");
  printf("0LO");
  for(i=0;i<=5;i++) printf("%9.4f",lp(im,i));
  im=findstm("LOC");
  printf("26400C");
  for(i=0;i<=5;i++) printf("%9.4f",lp(im,i));

  printf("0**************** End of SPERIL ********************0);
}


/********** find rule **********/
/*   return 1  :find*/
/*          -1 :EOF */
/*          0  :otherwise */
findr(fp,rsno)
  FILE *fp;
  char rsno[];       /*rule set no.*/
{
  int jj, k;
  char line[100], buf2[2];
  extern int pp;

  lget(fp,line);    /*one line get*/

  if((jj=line[0]) == EOF) k = -1;
  else
    {
     if(ccomp(line,"Rule",4) != 1) k=0;
     else

       {stmove(line,4,2,buf2);
        if(ccomp(buf2,rsno,2) == 1) k=1;
        else if(ccomp(buf2,"$$",2) == 1) k = -9;
        else k=0;
```

```
        }
    }
  if((pp=='y' || pp=='t') && k==1) printf("      %s", line);
  return(k);
}



/********** header classification **********/
/*   return type=1,2,3,4,5,6,7,8,9,10,11,12 */
header(line,pr)
  char line[];
  int  pr;      /*previous header type*/
{
  int  jj, k = -9;

  if      (ccomp(line,"      IF",7) == 1) k=1;
  else if(ccomp(line,"  OR IF",7) == 1) k=2;
  else if(ccomp(line,"THEN IF",7) == 1) k=3;
  else if(ccomp(line,"ELSE IF",7) == 1) k=4;

  else if(ccomp(line,"    and",7) == 1)
    {if(pr==1 || pr==2 || pr==5 || pr==6) k=5;   /*IF--ann*/
     else if(pr==3 || pr==4 || pr==7 || pr==8) k=7;
                                                  /*ELSE IF--and*/
     else if(pr==9 || pr==10) k=11;               /*THEN--and*/
    }
  else if(ccomp(line,"     or",7) == 1)
    {if(pr==1 || pr==2 || pr==5 || pr==6) k=6;  /*IF--or*/
     else if(pr==3 || pr==4 || pr==7 || pr==8) k=8;
                                                  /*ELSE IF--or*/
    }

  else if(ccomp(line,"   THEN",7) == 1) k=9;
  else if(ccomp(line,"   ELSE",7) == 1) k=10;

  else if(comp(line,"       ",7) != 1)
     { k = -10;
       printf("** undefined header -- miss writing of rule base **0); }

  return(k);
}



/********** get one line from file **********/
lget(fp,line)
  FILE *fp;
  char line[];
{
  int i=0, j;
  int c='a';
```

```
  while (c != '0)
    { c=getc(fp);           /*getc is in stdio.h */
      line[i]=c;
      i++; }
  line[i] = ' ';

  i++;
  if(i < 50)
    { for(j=0; j<=10;j++) line[i+j]=' '; }    /*insert 11 blank at tail*/
}


/********** string compare **********/
/*   return  1 if coincide */
comp(s,ss,n)
  char s[], ss[];
  int n;
{
  int i, k=1;

  for (i=0; i<n; i++)
    { if(s[i]!=ss[i] && s[i]!=' ' && ss[i]!=' ')
        { if(s[i]!='0 || s[i]!=' ')
            {k=0;  break; } }
    }
  return(k);
}

ccomp(s,ss,n)
   char s[], ss[];
{
   int i, k=1;

   for(i=0; i<n; i++)
      if(s[i] != ss[i]) {k=0; break; }
   return(k);
}


/********** string window move **********/
stmove(s,ist,n,ss)
  char s[], ss[];
  int ist, n;
{
  int i;
  for (i=0; i<n; i++)  ss[i]=s[ist+i];
}


/********** find short term memory **********/
/*   return  im (stm[im])     */
```

```
findstm(na)
  char na[];
{
  extern *smn[];
  int im, j=0, jj=0;

  while(jj != 1)
    { if((jj=comp(smn[j],na,3)) == 1)  im=j;
      else if((jj=comp(smn[j],"$$$",3)) ==1)  im = -1;
      j++;
    }
  if(im == -1)
     printf("  %s  is not in short term memory. -- error0, na);
  return(im);
}


/********** convert char[] to float **********/
float convert(buf)
  char buf[];
{
  float x=0.0, y=0.0;
  int   i, j, ii=0, sign=1;

  if (buf[0] == '-') { sign = -1;  ii=1; }

  for(i=ii; buf[i]>='0' && buf[i]<='9'; i++)
      x=10.0*x+(buf[i]-'0');

  if(buf[i] == '.')
    { for(j=i+1; buf[j]>='0' && buf[j]<='9'; j++)
          y=y+(buf[j]-'0')/(10.0*(j-i));
    }

  x=(x+y)*sign;
  return(x);
}


/********** min **********/
float min(x,y)
  float x, y;
{
  if(x>y) return(y);
  else return(x);
}

/********** max **********/
float max(x,y)
   float x, y;
{
```

```
    if(x>y) return(x);
    else return(y);
}

/********** printf sml[im] fo test ****************/
psml()
{
  extern char *sml[];
  int i;
  for(i=0;i<NSTM;i++)
    printf(" sml[%d]=%s",i,sml[i]);
  printf("0);
}
```

B6 Rule-base

(Demonstration purposes only)

```
Rule0101
      IF:GLO is   dest
   THEN:FIN dest 1
ELSE IF:GLO is   seve
   THEN:FIN seve 1
ELSE IF:GLO is   mode+slig
      or:LOC is   dest+seve
   THEN:FIN mode 1
ELSE IF:LOC is   mode+slig
   THEN:FIN slig 1
ELSE IF:GLO is   no
     and:LOC is   no
   THEN:FIN no   1
   ELSE:FIN uk

Rule0201
      IF:MAT is   r/c
THEN IF:STI is   dest
   THEN:GLO dest 0.6
ELSE IF:STI is   seve
   THEN:GLO seve 0.6
ELSE IF:STI is   mode
   THEN:GLO mode 0.6
ELSE IF:STI is   slig
   THEN:GLO slig 0.6
ELSE IF:STI is   no
   THEN:GLO no   0.6
   ELSE:GLO uk

Rule0202
      IF:MAT is   steel
THEN IF:STI is   dest
   THEN:GLO dest 0.8
ELSE IF:STI is   seve
   THEN:GLO seve 0.8
ELSE IF:STI is   mode
   THEN:GLO mode 0.8
ELSE IF:STI is   slig
   THEN:GLO slig 0.8
ELSE IF:STI is   no
   THEN:GLO no   0.8
   ELSE:GLO uk

Rule0203
      IF:DAM is   dest
   THEN:GLO dest 0.4
ELSE IF:DAM is   seve
   THEN:GLO seve 0.4
ELSE IF:DAM is   mode
   THEN:GLO mode 0.4
ELSE IF:DAM is   slig
   THEN:GLO slig 0.4
ELSE IF:DAM is   no
```

```
   THEN:GLO no   0.4
   ELSE:GLO uk

Rule0204
      IF:DRI is   dest
   THEN:GLO dest 0.6
ELSE IF:DRI is   seve
   THEN:GLO seve 0.6
ELSE IF:DRI is   mode
   THEN:GLO mode 0.6
ELSE IF:DRI is   slig
   THEN:GLO slig 0.6
ELSE IF:DRI is   no
   THEN:GLO no   0.6
   ELSE:GLO uk

Rule0205
      IF:INS is   careful
THEN IF:VST is   dest
   THEN:GLO dest 1
ELSE IF:VST is   seve
   THEN:GLO seve 1
ELSE IF:VST is   mode
   THEN:GLO mode 0.9
ELSE IF:VST is   slig
   THEN:GLO slig 0.8
ELSE IF:VST is   no
   THEN:GLO no   0.8
   ELSE:GLO uk

Rule0206
      IF:INS is   ml=more-or-less
THEN IF:VST is   dest
   THEN:GLO dest 1
ELSE IF:VST is   seve
   THEN:GLO seve 1
ELSE IF:VST is   mode
   THEN:GLO mode 0.9
ELSE IF:VST is   slig
   THEN:GLO slig 0.7
ELSE IF:VST is   no
   THEN:GLO no   0.6
   ELSE:GLO uk

Rule0207
      IF:INS is   rough
      or:INS is   uk
THEN IF:VST is   dest
   THEN:GLO dest 1
ELSE IF:VST is   seve
   THEN:GLO seve 1
ELSE IF:VST is   mode
   THEN:GLO mode 0.8
```

```
ELSE IF:VST is   slig
   THEN:GLO slig 0.6
ELSE IF:VST is   no
   THEN:GLO no   0.4
   ELSE:GLO uk

Rule0401
     IF:VNS is   dest
   THEN:LOC dest 0.7
ELSE IF:VNS is   seve
   THEN:LOC seve 0.7
ESLE IF:VNS is   mode
   THEN:LOC mode 0.7
ELSE IF:VNS is   slig
   THEN:LOC slig 0.7
ELSE IF:VNS is   no
   THEN:LOC no   0.7
   ELSE:LOC uk

Rule0402
     IF:DRI is   dest
   THEN:LOC dest 0.7
ELSE IF:DRI is   seve
   THEN:LOC seve 0.7
ELSE IF:DRI is   mode
   THEN:LOC mode 0.7
ELSE IF:DRI is   slig
   THEN:LOC slig 0.7
ELSE IF:DRI is   no
   THEN:LOC no   0.7
   ELSE:LOC uk

Rule0501
     IF:MAT is   r/c
THEN IF:ISD <=   -8.9
   THEN:DRI uk   1
ELSE IF:ISD <=   0.4
   THEN:DRI no   0.9
ELSE IF:ISD <=   0.8
   THEN:DRI slig 0.9
ELSE IF:ISD <=   1.3
   THEN:DRI mode 0.9
ELSE IF:ISD <=   2.0
   THEN:DRI seve 0.9
ELSE IF:ISD >    2.0
   THEN:DRI dest 0.9
   ELSE:DRI uk

Rule0502
     IF:MAT is   r/c
   THEN:FLX =    no

Rule0503
```

```
      IF:MAT is   steel
     and:VST >=   8
     and:FLX is   yes
THEN IF:ISD <=   0.6
   THEN:DRI no   0.8
ELSE IF:ISD <=   1.2
   THEN:DRI slig 0.8
ELSE IF:ISD <=   1.9
   THEN:DRI mode 0.8
ELSE IF:ISD <=   3.0
   THEN:DRI seve 0.8
ELSE IF:ISD >    3.0
   THEN:DRI dest 0.8
   ELSE:DRI uk

Rule0504
      IF:NST <    8
      or:FLX is   no  +uk
     and:MAT is   steel
THEN IF:ISD <=   0.4
   THEN:DRI no   0.8
ELSE IF:ISD <=   0.9
   THEN:DRI slig 0.8
ELSE IF:ISD <=   1.5
   THEN:DRI mode 0.8
ELSE IF:ISD <=   2.3
   THEN:DRI seve 0.8
ELSE IF:ISD >    2.3
   THEN:DRI dest 0.8
   ELSE:DRI uk

Rule0601
      IF:MAT is    r/c
THEN IF:NST <=    4
   THEN:IFS =     2.0  (8.0/NST)
ELSE IF:NST <=    6
   THEN:IFS =     1.33
ELSE IF:NST <=    8
   THEN:IFS =     1.0
ELSE IF:NST <=    11
   THEN:IFS =     0.72
ELSE IF:NST <=    14
   THEN:IFS =     0.57
ELSE IF:NST <=    18
   THEN:IFS =     0.44
ELSE IF:NST <=    23
   THEN:IFS =     0.32
ELSE IF:NST <=    30
   THEN:IFS =     0.27
ELSE IF:NST <=    40
   THEN:IFS =     0.2
ELSE IF:NST >     40
   THEN:IFS =     0.17
```

```
Rule0602
      IF:MAT  is    steel
THEN IF:NST  <=    4
   THEN:IFS  =     1.62  (6.5/NST)
ELSE IF:NST  <=    6
   THEN:IFS  =     1.1
ELSE IF:NST  <=    8
   THEN:IFS  =     0.81
ELSE IF:NST  <=    11
   THEN:IFS  =     0.59
ELSE IF:NST  <=    14
   THEN:IFS  =     0.46
ELSE IF:NST  <=    18
   THEN:IFS  =     0.36
ELSE IF:NST  <=    23
   THEN:IFS  =     0.28
ELSE IF:NST  <=    30
   THEN:IFS  =     0.22
ELSE IF:NST  <=    40
   THEN:IFS  =     0.163
ELSE IF:NST  <=    55
   THEN:IFS  =     0.118
ESLE IF:NST  <=    70
   THEN:IFS  =     0.093
ELSE IF:NST  >     70
   THEN:IFS  =     0.075

Rule0603
      IF:MAT  is    r/c
     and:INF  >=    IFS
THEN IF:CNF  <     -8.9
   THEN:STI  uk    1
ELSE IF:CNF  <=    4
   THEN:STI  no    0.9
ELSE IF:CNF  <=    12
   THEN:STI  slig  0.9
ELSE IF:CNF  <=    22
   THEN:STI  mode  0.9
ELSE IF:CNF  <=    40
   THEN:STI  seve  0.9
ELSE IF:CNF  >     40
   THEN:STI  dest  0.9
   ELSE:STI  uk

Rule0604
      IF:MAT  is    r/c
     and:INF  <     IFS
THEN IF:CNF  <     -8.9
   THEN:STI  uk    1
ELSE IF:CNF  <=    6
   THEN:STI  slig  0.9
ELSE IF:CNF  <=    15
   THEN:STI  mode  0.9
```

```
ELSE IF:CNF <=   30
   THEN:STI seve 0.9
ELSE IF:CNF >    30
   THEN:STI dest 0.9
   ELSE:STI uk

Rule0605
      IF:MAT is   steel
     and:INF >=   IFS
THEN IF:CNF <    -8.9
   THEN:STI uk   1
ELSE IF:CNF <=   4
   THEN:STI no   0.9
ELSE IF:CNF <=   12
   THEN:STI slig 0.9
ELSE IF:CNF <=   22
   THEN:STI mode 0.9
ELSE IF:CNF <=   40
   THEN:STI seve 0.9
ELSE IF:CNF >    40
   THEN:STI dest 0.9
   ELSE:STI uk

Rule0606
      IF:MAT is   steel
     and:INF <    IFS
THEN IF:CNF <    -8.9
   THEN:STI uk   1
ELSE IF:CNF <=   6
   THEN:STI slig 0.9
ELSE IF:CNF <=   15
   THEN:STI mode 0.9
ELSE IF:CNF <=   30
   THEN:STI seve 0.9
ELSE IF:CNF >    30
   THEN:STI dest 0.9
   ELSE:STI uk

Rule0701
      IF:CDI <    -8.9
      or:CDD <    -8.9
   THEN:DAM uk   1
ELSE IF:CDD <    3   (no decrease)
     and:CDI <=   10
   THEN:DAM no   0.8
ELSE IF:CDD <    3
     and:CDI <=   40
   THEN:DAM slig 0.8
ELSE IF:CDD <=   15
   THEN:DAM mode 0.8
ELSE IF:CDD <=   40
   THEN:DAM seve 0.8
ELSE IF:CDD >    40
```

```
   THEN:DAM dest 0.8
   ELSE:DAM uk

Rule0901
     IF:MAT is   steel
THEN IF:S01 is   yes     (partial collapse)
   THEN:VST dest 1
ELSE IF:S02 is   yes     (buckling of column)
   THEN:VST dest 0.5
    and:VST seve 0.5
ELSE IF:S03 is   yes     (buckling of girder/beam)
     or:S04 is   yes     (buckling of diagonal bracing)
     or:S05 is   yes     (deformation or loosening of joint)
   THEN:VST seve 0.9
ELSE IF:S06 is   yes    (spalling/crack on shear wall)
   THEN:VST mode 0.8
ELSE IF:S07 is   yes    (spalling/crack on exterior/interior wall)
     or:S08 is   yes    (spalling/crack on floor)
   THEN:VST mode 0.5
    and:VST slig 0.5
ELSE IF:S01 is   no
    and:S02 is   no
    and:S03 is   no
    and:S04 is   no
    and:S05 is   no
    and:S06 is   no
    and:S07 is   no
    and:S08 is   no
   THEN:VST no   1
   ELSE:VST uk

Rule0902
     IF:MAT is   r/c
THEN IF:C01 is   yes    (partial collapse)
   THEN:VST dest 1
ELSE IF:C02 is   yes    (large spalling on column)
   THEN:VST dest 0.5
    and:VST seve 0.5
ELSE IF:C03 is   yes    (large spa. on load-bear./shear wall or girder/beam)
   THEN:VST seve 0.8
ELSE IF:C04 is   yes    (small spa. on load-bear./shear wall or girder/beam)
   THEN:VST seve 0.5
    and:VST mode 0.5
ELSE IF:C05 is   yes    (small cracks on load-bear./shear wall or girder/beam)
     or:C06 is   yes    (spalling or large cracks on outer walls or floor)
   THEN:VST mode 0.5
    and:VST slig 0.5
ELSE IF:C01 is   no
    and:C02 is   no
    and:C03 is   no
    and:C04 is   no
    and:C05 is   no
    and:C06 is   no
```

```
   THEN:VST no    1
   ELSE:VST uk

Rule1001
     IF:N01 is    severe    (nonstructural partition)
   THEN:SUM ++    10
ELSE IF:N01 is    considerable
   THEN:SUM ++    5
ELSE IF:N01 is    slight
   THEN:SUM ++    2.5

Rule1002
     IF:N02 is    severe   (windows)
   THEN:SUM ++    10
ELSE IF:N02 is    considerable
   THEN:SUM ++    5
ELSE IF:N02 is    slight
   THEN:SUM ++    2.5

Rule1003
     IF:N03 is    severe   (doors)
   THEN:SUM ++    10
ELSE IF:N03 is    considerable
   THEN:SUM ++    5
ELSE IF:N03 is    slight
   THEN:SUM ++    2.5

Rule1004
     IF:N04 is    severe   (ceiling and light fixtures)
   THEN:SUM ++    10
ELSE IF:N04 is    considerable
   THEN:SUM ++    5
ELSE IF:N04 is    slight
   THEN:SUM ++    2.5

Rule1005
     IF:N05 is    severe   (stairs)
   THEN:SUM ++    10
ELSE IF:N05 is    considerable
   THEN:SUM ++    5
ELSE IF:N05 is    slight
   THEN:SUM ++    2.5

Rule1006
     IF:N06 is    severe   (elevator)
   THEN:SUM ++    10
ELSE IF:N06 is    considerable
   THEN:SUM ++    5
ELSE IF:N06 is    slight
   THEN:SUM ++    2.5

Rule1007
     IF:N07 is    severe   (electricity)
```

```
   THEN:SUM ++    10
ELSE IF:N07 is    considerable
   THEN:SUM ++    5
ELSE IF:N07 is    slight
   THEN:SUM ++    2.5

Rule1008
     IF:N08 is    severe   (air coditioning)
   THEN:SUM ++    10
ELSE IF:N08 is    considerable
   THEN:SUM ++    5
ELSE IF:N08 is    slight
   THEN:SUM ++    2.5

Rule1009
     IF:N09 is    severe   (water works)
   THEN:SUM ++    10
ELSE IF:N09 is    considerable
   THEN:SUM ++    5
ELSE IF:N09 is    slight
   THEN:SUM ++    2.5

Rule1010
     IF:N10 is    severe   (gas facility)
   THEN:SUM ++    10
ELSE IF:N10 is    considerable
   THEN:SUM ++    5
ELSE IF:N10 is    slight
   THEN:SUM ++    2.5

Rule1011
     IF:N11 is    severe   (communication)
   THEN:SUM ++    10
ELSE IF:N11 is    considerable
   THEN:SUM ++    5
ELSE IF:N11 is    slight
   THEN:SUM ++    2.5

Rule1012
     IF:SUM <=    3
   THEN:VNT no    0.8
ELSE IF:SUM <=    15
   THEN:VNT slig 0.8
ELSE IF:SUM <=    35
   THEN:VNT mode 0.8
ELSE IF:SUM <=    60
   THEN:VNT seve 0.9
ELSE IF:SUM >     60
   THEN:VNT dest 1
   ELSE:VNT uk

Rule$$$$       mark for end of rules *****************
```

```
dest : destructive
seve : severe
mode : moderate
slig : slight
no   : no
uk   : unknown
r/c  : reinforced concrete

FIN  : final goal -- damage state
GLO  : damage of global nature
CUM  : damage of cumulative nature
LOC  : damage of local nature
DRI  : damage due to drifting
STI  : damage of stiffness
DAM  : damage of damping
FAT  : fatigue & creep
VST  : visual damage of structural member
VNT  : visual damage of nonstructural member
MAT  : material of structure
ISD  : interstory drift
CNF  : change of natural frequency of vibration
INF  : initial natural frequency
IFS  : reference of natural frequency
CDI  : change of damping -- increase
CDD  : change of damping -- decrease
FLX  : flexible design
INS  : careful visual inspection
SUM  : buffer memory for calculation of VNS

S01  : check items of visual structural damage for steel
 |
S07
C01  : check items of visual structural damage for R/C
 |
C06
N01  : check items of visual nonstructural damage
 |
N11
```

B7 Question list

(Demonstration purposes only)

```
Q-MAT
What is the material of the building { r/c(reinforced concrete), steel }?
Q-INS
Was the visual inspection done carefully?  Did you check even inside
the covers?   { care(careful), ml(more or less), roug(rough), uk(unknown) }
Q-FLX
Is the building designed to be flexible?  { y(yes), n(no), uk(unknown) }
Q-ISD
What was the maximum interstory drift during the earthquake?
{ drift/height*100 [%] ,  -9(unknown) }
Q-NST
Enter the number of stories.
Q-INF
What was the initial natural frequency [Hz] of the vibration?  { -9(unknown) }
Q-CNF
What was the decrease [%] of the natural frequency during the eathquake?
 { -9(unknown) }
Q-CDI
What was the increase [%] of damping?  { 0(almost no increase), -9(unknown)
Q-CDD
What was the decrease [%] of damping from its peak value?
If no initial increase process was observed, use initial value instead
of the peak value.   { 0(almost no increase), -9(unknown) }
Q-S01
Is there partial collapse observed?  { y(yes), g(grey), n(no), uk(unknown) }
Q-S02
Is there the buckling of column? { y, g, n, uk }
Q-S03
Is there the buckling of girder/beam?  { y, g, n, uk }
Q-S04
Is there the buckling of diagonal bracing?  { y, g, n, uk }
Q-S05
Is there the deformation or loosening of joint?  { y, g, n, uk }
Q-S06
Are there considerable spallings/cracks observed on shear wall?
 { y(yes), g(grey), n(no), uk(unknown) }
Q-S07
Are there considerable spallings/cracks observed on other exterior/
interior walls? { y, g, n, uk }
Q-S08
Are there considerable spallings/cracks observed on floors?  { y, g, n, uk }
Q-C01
Is there partial collapse observed?  { y(yes), g(grey), n(no), uk(unknown) }
Q-C02
Is there large spalling observed on column?  { y, g, n, uk }
Q-C03
Is there large spalling observed on load bearing wall, shear wall, or
girder/beam?  { y, g, n, uk }
Q-C04
Are there small spallings or large cracks observed on column,
load bearing wall or girder/beam?  { y, g, n, uk }
Q-C05
Are there considerable number of small cracks observed on column, load
```

```
bearing wall or girder/beam?  { y, g, n, uk }
Q-C06
Are there spallings or large cracks observed on other walls or floors?
  { y, g, n, uk }
Q-N01
How is the damage of nonstructural partitions?
 { se(severe), co(considerable) sl(slight), no , uk(unknown) }
 {     severe      : more than 10 %     }
 {     considerable: approximately 5 % }
Q-N02
How is the damage of windows?  { se, co, sl, no, uk }
Q-N03
How is the damage of doors?  { se, co, sl, no, uk }
Q-N04
How is the damage of ceilings including light fixtures?
  { se, co, sl, uk }
Q-N05
How is the damage of stairs? { se, co, sl, no, uk }
Q-N06
How is the damage of elevators?
  { se(severe), co(considerable), sl(slight), no , uk(unknown) }
Q-N07
How is the damage of electricity?  { se, co, sl, no, uk }
Q-N08
How is the damage of air conditioning?  { se, co, sl, no, uk }
Q-N09
How is the damage of waterworks?  { se, co, sl, no, uk }
Q-N10
How is the damage of gas facility?  { se, co, sl, no, uk }
Q-N11
How is the damage of communication facility?  { se, co, sl, no, uk }
Q-$$$     this is the end mark of file-quest accessed by speril.
```

B8 Typical runs

(Demonstration purposes only)

```
% speril
****************** SPERIL  (Version I) ********************
 Rule-based Damage Assessment System for Existing Structure
 subjected to Earthquake Excitation
***********************************************************
Answer the following questions.
Then I can suggest a reasonable answer.

Enter the name of building. =shelard
Enter date. = 1975
Do you want to see rule sentences on display {y or n} ? =y
     Rule0501
          IF:MAT is    r/c
What is the material of the building { r/c(reinforced concrete), steel }?
===r/c
     THEN IF:ISD <=    -8.9
What was the maximum interstory drift during the earthquake?
{ drift/height*100 [%] ,  -9(unknown) }
===-9
        THEN:DRI uk    1
     DRI :  0.00       0.00       0.00       0.00       0.00       1.00
            0.00 0.00 0.00 0.00 0.00 0.00 0.00 0.00 0.00 1.00

     Rule0502
          IF:MAT is    r/c
        THEN:FLX =     no
     FLX : no

     Rule0503
          IF:MAT is    steel
         and:NST >=    8
Enter the number of stories.
===5
         and:FLX is    yes
        ELSE:DRI uk
     Rule0504
          IF:NST <     8
          or:FLX is    no  +uk
         and:MAT is    steel
        ELSE:DRI uk
     Rule0601
          IF:MAT is    r/c
     THEN IF:NST <=    4
        THEN:IFS =     2.0  (8.0/NST)
     ELSE IF:NST <=    6
        THEN:IFS =     1.33
     IFS :     1.450
               1.450

     Rule0602
          IF:MAT is    steel
     Rule0603
          IF:MAT is    r/c
```

```
          and:INF >=    IFS
What was the initial natural frequency [Hz] of the vibration?  { -9(unknown) }
===-9
        ELSE:STI uk
     Rule0604
           IF:MAT is    r/c
          and:INF <     IFS
     THEN IF:CNF <      -8.9
What was the decrease [%] of the natural frequency during the eathquake?
 { -9(unknown) }
===-9
        THEN:STI uk     1
     STI :  0.00        0.00        0.00        0.00        0.00        1.00
            0.00 0.00 0.00 0.00 0.00 0.00 0.00 0.00 0.00 1.00

     Rule0605
           IF:MAT is    steel
          and:INF >=    IFS
        ELSE:STI uk
     Rule0606
           IF:MAT is    steel
          and:INF <     IFS
        ELSE:STI uk
     Rule0701
           IF:CDI <     -8.9
What was the increase [%] of damping?  { 0(almost no increase), -9(unknown)
===-9
           or:CDD <     -8.9
What was the decrease [%] of damping from its peak value?
If no initial increase process was observed, use initial value instead
of the peak value.    { 0(almost no increase), -9(unknown) }
===-9
        THEN:DAM uk     1
     DAM :  0.00        0.00        0.00        0.00        0.00        1.00
            0.00 0.00 0.00 0.00 0.00 0.00 0.00 0.00 0.00 1.00

     Rule0901
           IF:MAT is    steel
        ELSE:VST uk
     Rule0902
           IF:MAT is    r/c
     THEN IF:C01 is     yes     (partial collapse)
Is there partial collapse observed?  { y(yes), g(grey), n(no), uk(unknown) }
===y
        THEN:VST dest 1
     VST :  0.00        0.00        0.00        0.00        1.00        0.00
            0.00 0.00 0.00 0.00 0.00 0.00 0.00 0.00 1.00 0.00

     Rule1001
           IF:N01 is    severe     (nonstructural partition)
How is the damage of nonstructural partitions?
 { se(severe), co(considerable) sl(slight), no , uk(unknown) }
 {     severe       : more than 10 %     }
```

```
 {    considerable: approximately 5 % }
===sl
        THEN:SUM ++    10
     ELSE IF:N01 is    considerable
        THEN:SUM ++    5
     ELSE IF:N01 is    slight
        THEN:SUM ++    2.5
     SUM :      2.500
                2.500

     Rule1002
          IF:N02 is    severe    (windows)
How is the damage of windows?  { se, co, sl, no, uk }
===no
        THEN:SUM ++    10
     ELSE IF:N02 is    considerable
        THEN:SUM ++    5
     ELSE IF:N02 is    slight
        THEN:SUM ++    2.5
     Rule1003
          IF:N03 is    severe    (doors)
How is the damage of doors?  { se, co, sl, no, uk }
===no
        THEN:SUM ++    10
     ELSE IF:N03 is    considerable
        THEN:SUM ++    5
     ELSE IF:N03 is    slight
        THEN:SUM ++    2.5
     Rule1004
          IF:N04 is    severe    (ceiling and light fixtures)
How is the damage of ceilings including light fixtures?
  { se, co, sl, no, uk }
===uk
        THEN:SUM ++    10
     ELSE IF:N04 is    considerable
        THEN:SUM ++    5
     ELSE IF:N04 is    slight
        THEN:SUM ++    2.5
     Rule1005
          IF:N05 is    severe    (stairs)
How is the damage of stairs? { se, co, sl, no, uk }
===no
        THEN:SUM ++    10
     ELSE IF:N05 is    considerable
        THEN:SUM ++    5
     ELSE IF:N05 is    slight
        THEN:SUM ++    2.5
     Rule1006
          IF:N06 is    severe    (elevator)
How is the damage of elevators?
  { se(severe), co(considerable), sl(slight), no , uk(unknown) }
===no
        THEN:SUM ++    10
```

```
      ELSE IF:N06 is    considerable
         THEN:SUM ++    5
      ELSE IF:N06 is    slight
         THEN:SUM ++    2.5
      Rule1007
           IF:N07 is    severe   (electricity)
How is the damage of electricity?  { se, co, sl, no, uk }
===no
         THEN:SUM ++    10
      ELSE IF:N07 is    considerable
         THEN:SUM ++    5
      ELSE IF:N07 is    slight
         THEN:SUM ++    2.5
      Rule1008
           IF:N08 is    severe   (air coditioning)
How is the damage of air conditioning?  { se, co, sl, no, uk }
===no
         THEN:SUM ++    10
      ELSE IF:N08 is    considerable
         THEN:SUM ++    5
      ELSE IF:N08 is    slight
         THEN:SUM ++    2.5
      Rule1009
           IF:N09 is    severe   (water works)
How is the damage of waterworks?  { se, co, sl, no, uk }
===no
         THEN:SUM ++    10
      ELSE IF:N09 is    considerable
         THEN:SUM ++    5
      ELSE IF:N09 is    slight
         THEN:SUM ++    2.5
      Rule1010
           IF:N10 is    severe   (gas facility)
How is the damage of gas facility?  { se, co, sl, no, uk }
===no
         THEN:SUM ++    10
      ELSE IF:N10 is    considerable
         THEN:SUM ++    5
      ELSE IF:N10 is    slight
         THEN:SUM ++    2.5
      Rule1011
           IF:N11 is    severe   (communication)
How is the damage of communication facility?  { se, co, sl, no, uk }
===no
         THEN:SUM ++    10
      ELSE IF:N11 is    considerable
         THEN:SUM ++    5
      ELSE IF:N11 is    slight
         THEN:SUM ++    2.5
      Rule1012
           IF:SUM <=    3
         THEN:VNS no    0.8
      VNS :  0.80       0.00       0.00       0.00       0.00       0.20
```

```
          0.80 0.00 0.00 0.00 0.00 0.00 0.00 0.00 0.00 0.20

      Rule0201
           IF:MAT is   r/c
      THEN IF:STI is   dest
         THEN:GLO dest 0.6
      ELSE IF:STI is   seve
         THEN:GLO seve 0.6
      ELSE IF:STI is   mode
         THEN:GLO mode 0.6
      ELSE IF:STI is   slig
         THEN:GLO slig 0.6
      ELSE IF:STI is   no
         THEN:GLO no   0.6
         ELSE:GLO uk
      Rule0202
           IF:MAT is   steel
         ELSE:GLO uk
      Rule0203
           IF:DAM is   dest
         THEN:GLO dest 0.4
      ELSE IF:DAM is   seve
         THEN:GLO seve 0.4
      ELSE IF:DAM is   mode
         THEN:GLO mode 0.4
      ELSE IF:DAM is   slig
         THEN:GLO slig 0.4
      ELSE IF:DAM is   no
         THEN:GLO no   0.4
         ELSE:GLO uk
      Rule0204
           IF:DRI is   dest
         THEN:GLO dest 0.6
      ELSE IF:DRI is   seve
         THEN:GLO seve 0.6
      ELSE IF:DRI is   mode
         THEN:GLO mode 0.6
      ELSE IF:DRI is   slig
         THEN:GLO slig 0.6
      ELSE IF:DRI is   no
         THEN:GLO no   0.6
         ELSE:GLO uk
      Rule0205
           IF:INS is   careful
Was the visual inspection done craefully?  Did you check even inside
the covers?  { care(careful), ml(more or less), roug(rough), uk(unknown) }
===care
      THEN IF:VST is   dest
         THEN:GLO dest 1
      GLO :  0.00      0.00      0.00      0.00      1.00      0.00
             0.00 0.00 0.00 0.00 0.00 0.00 0.00 0.00 1.00 0.00

      Rule0206
```

```
          IF:INS is   ml=more-or-less
        ELSE:GLO uk
     Rule0207
          IF:INS is   rough
          or:INS is   uk
        ELSE:GLO uk
     Rule0401
          IF:VNS is   dest
        THEN:LOC dest 0.7
     ELSE IF:VNS is   seve
        THEN:LOC seve 0.7
     Rule0402
          IF:DRI is   dest
        THEN:LOC dest 0.7
     ELSE IF:DRI is   seve
        THEN:LOC seve 0.7
     ELSE IF:DRI is   mode
        THEN:LOC mode 0.7
     ELSE IF:DRI is   slig
        THEN:LOC slig 0.7
     ELSE IF:DRI is   no
        THEN:LOC no   0.7
        ELSE:LOC uk
     Rule0101
          IF:GLO is   dest
        THEN:FIN dest 1
     FIN :  0.00      0.00      0.00      0.00      1.00      0.00
            0.00 0.00 0.00 0.00 0.00 0.00 0.00 0.00 1.00 0.00


****************************************************

  The damage is destructive.

*****************************************************

          no    slight moderate   severe destruct   unknown
FIN   0.0000   0.0000   0.0000   0.0000   1.0000   0.0000
GLO   0.0000   0.0000   0.0000   0.0000   1.0000   0.0000
LOC   0.0000   0.0000   0.0000   0.0000   0.0000   0.0000

***************** End of SPERIL ********************
%
%
%
%
%
%
% speril
****************** SPERIL  (Version I) ********************
 Rule-based Damage Assessment System for Existing Structure
 subjected to Earthquake Excitation
```

```
****************************************************************
Answer the following questions.
Then I can suggest a reasonable answer.

Enter the name of building. =mprc
Enter date. = 1974
Do you want to see rule sentences on display {y or n} ? =n
What is the material of the building { r/c(reinforced concrete), steel }?
===r/c
What was the maximum interstory drift during the earthquake?
{ drift/height*100 [%] ,  -9(unknown) }
===-9
Enter the number of stories.
===6
What was the initial natural frequency [Hz] of the vibration?  { -9(unknown ) }
===-9
What was the decrease [%] of the natural frequency during the eathquake?
 { -9(unknown) }
===-9
What was the increase [%] of damping?  { 0(almost no increase), -9(unknown) }
===-9
What was the decrease [%] of damping from its peak value?
If no initial increase process was observed, use initial value instead
of the peak value.   { 0(almost no increase), -9(unknown) }
===-9
Is there partial collapse observed?  { y(yes), g(grey), n(no), uk(unknown) }
===y
How is the damage of nonstructural partitions?
 { se(severe), co(considerable) sl(slight), no , uk(unknown) }
 {     severe      : more than 10 %     }
 {     considerable: approximately 5 % }
===se
How is the damage of windows?  { se, co, sl, no, uk }
===se
How is the damage of doors?  { se, co, sl, no, uk }
===se
How is the damage of ceilings including light fixtures?
  { se, co, sl, no, uk }
===se
How is the damage of stairs? { se, co, sl, no, uk }
===se
How is the damage of elevators?
  { se(severe), co(considerable), sl(slight), no , uk(unknown) }
===se
How is the damage of electricity?  { se, co, sl, no, uk }
===se
How is the damage of air conditioning?  { se, co, sl, no, uk }
===no
How is the damage of waterworks?  { se, co, sl, no, uk }
===no
How is the damage of gas facility?  { se, co, sl, no, uk }
===no
How is the damage of communication facility?  { se, co, sl, no, uk }
```

```
===se
Was the visual inspection done craefully?  Did you check even inside
the covers?    { care(careful), ml(more or less), roug(rough), uk(unknown) }
===care


*******************************************************

  The damage is destructive.

*******************************************************

          no      slight  moderate    severe  destruct    unknown
FIN   0.0000    0.0000    0.0000    0.0000    1.0000    0.0000
GLO   0.0000    0.0000    0.0000    0.0000    1.0000    0.0000
LOC   0.0000    0.0000    0.0000    0.0000    0.7000    0.3000

***************** End of SPERIL ********************
%
```

References

1. Abrams, D. P., and Sozen, M. A., Experimental study of frame-wall interaction in reinforced concrete structures subjected to strong earthquake motions, *Structural Research Series*, No. 460, Department of Civil Engineering, University of Illinois, Urbana, IL, May 1979.
2. Allen, D. E., Limit states design—a probabilistic study, *Canadian Journal of Civil Engineering*, **2,** No. 1, 1975, 36–49.
3. Andrews, H. C., *Introduction to Mathematical Techniques in Pattern Recognition*, Wiley-Interscience, New York, 1972.
4. Ang, A. H.-S. (Chairman, Task Committee on Structural Safety), Structural safety—a literature review, *Journal of the Structural Engineering Division, ASCE*, **98,** No. ST4, April 1972, 845–884.
5. Ang, A. H.-S., and Cornell, C. A., Reliability bases of structural safety and design, *Journal of the Structural Division, American Society of Civil Engineers,* **100,** No. ST9, September 1974, 1755–1769.
6. Aristizabal-Ochoa, J. D., and Sozen, M. A., Behavior of ten-story reinforced concrete walls subjected to earthquake motions, *Structural Research Series*, No. 431, Department of Civil Engineering, University of Illinois, Urbana, IL, October 1976.
7. Baldwin, J. W., Jr., Salane, H. J., and Duffield, R. C., Fatigue test of a three-span composite highway bridge, *Study 71-1*, Department of Civil Engineering, University of Missouri-Columbia, June 1978.
8. Beck, J. L., *Determining models of structures from earthquake records*, Report No. EERL 78-01, California Institute of Technology, Pasadena, CA, June 1978.
9. Beck, J. L., and Jennings, P. C., Structural identification using linear models and earthquake records (private communications, 1978).
10. Berens, A. P., Gallagher, J. P., and Hovey, P. W., Statistics of crack growth, in: *Proceedings Ninth U.S. National Congress of Applied Mechanics*, ASME, 1982, 325–328.
11. Bertero, V. V., and Bresler, B., Design and engineering decisions: failure criteria (limit states), in: *Developing methodologies for evaluating the earthquake safety of existing buildings*, Report No. UCB-EERC-77/06, Earthquake Engineering Research Center, University of California at Berkeley, February 1977, 114–142.
12. Blejwas, T., and Bresler, B., *Damageability in existing buildings*, Report No. UCB/EERC-78/12, Earthquake Engineering Research Center, University of California at Berkeley, August 1979.
13. Blume, J. A., and Monroe, R. E., *The spectral matrix method of predicting damage from ground motion*, Report No. JAP-99-88, John Blume & Associates, 1971.

14. Blockley, D. I., Predicting the likelihood of structural accidents, *Proceedings Institution of Civil Engineers,* **59,** Part 2, December 1975, 659–668.
15. Bresler, B., Evaluation of earthquake safety of existing buildings, *Developing methodologies for evaluating the earthquake safety of existing buildings*, Report No. UCB/EERC-77/06, Earthquake Engineering Research Center, University of California at Berkeley, February 1977, 1–15.
16. Bresler, B., Behavior of structural elements—a review, *Building Practices for Disaster Mitigation*, Edited by R. Wright, S. Kramer, and C. Culver, National Bureau of Standards, Building Science Series No. 46, February 1973, 286–351.
17. Bresler, B., and Hanson, J. M., Damageability and reliability of existing structures, in: *Proceedings Ninth U.S. National Congress of Applied Mechanics*, ASME, 1982, 309–313.
18. Bresler, B., Hanson, J. M., Comartin, C. D., and Thomasen, S. E., *Practical Evaluation of Structural Reliability*, Preprint 80-596, ASCE Convention, Florida, October 27–31, 1980.
19. Bresler, B., Okada, T., and Zisling, D., Assessment of earthquake safety and of hazard abatement, in: *Developing methodologies for evaluating the earthquake safety of existing buildings*, Report No. UCB/EERC-77/06, Earthquake Engineering Research Center, University of California at Berkeley, February 1977, 17–49.
20. Brown, C. B., A fuzzy safety measure, *Journal of the Engineering Mechanics Division,* **105,** No. EM5, October 1979, 855–872.
21. Brown, C. B., Analytical methods for environmental assessment and decision making, in: *Regional Environmental Systems*, NSF/ENV76-04273, Department of Civil Engineering, University of Washington, Seattle, WA, June 1977, 180–206.
22. Brown, C. B., and Leonard, R. S., *Subjective uncertainty analysis*, Preprint No. 1388, ASCE National Structural Engineering Meeting, Baltimore, Maryland, 19–23 April 1971.
23. Brown, C. B., and Yao, J. T. P., Fuzzy sets in structural engineering, *Journal of the Structural Engineering Division, ASCE*, 1983.
24. Cecen, H., Responses of ten-story reinforced concrete frames to simulated earthquakes, Ph.D. Thesis, School of Civil Engineering, University of Illinois, Urbana, IL, March 1979.
25. Chen, S. J. H., Methods of system identification in structural engineering, M.S. Thesis, School of Civil Engineering, Purdue University, West Lafayette, IN 47907, August 1976.
26. Chen, S. J. H., System identification and damage assessment of existing structures, Ph.D. Thesis, School of Civil Engineering, Purdue University, West Lafayette, IN 47907, December 1980.
27. Chen, S. J. H., and Yao, J. T. P., Identification of structural damage using earthquake response data, in: *Proceedings of the ASCE EMD-STD Specialty Conference*, Austin, Texas, September 17–19, 1979, 661–664.
28. Chen, S. J. H., and Yao, J. T. P., *Data analyses for safety evaluation of existing structures*, Technical Report No. CE-STR-80-18, School of Civil Engineering, Purdue University, W. Lafayette, IN, December 1980.
29. Chou, I.-H., Decision theory in earthquake engineering, Ph.D. Thesis, School of Civil Engineering, Purdue University, West Lafayette, IN, December 1973.
30. Collins, J. D., Hart, G. C., Hasselman, T. K., and Kennedy, B., Statistical identification of structures, *AIAA Journal,* **12,** 1970, 185–190.
31. Collins, J. D., Hart, G. C., and Kennedy, B., Statistical analysis of the modal

properties of large structural systems, Paper No. 71078S, in: *SAE National Aeronautic and Space Engineering and Manufacturing Meeting*, Los Angeles, Calif., September 28–30, 1971.
32. Collins, J. D., and Thomson, W. T., The eigenvalue problem for structural systems with statistical properties, *AIAA Journal,* **7,** No. 4, April 1969, 642–648.
33. Collins, J. D., Young, J. P., and Keifling, L. A., Methods and applications of system identification in shock and vibration, in: *System Identification of Vibrating Structures*, Edited by W. Pilky and R. Cohen, ASME, New York, 1972, 45–72.
34. Comité Européen du Béton, *Code Modèle pour les Structures en Béton (Volume 11—International System of Unified Standard Codes of Practice for Structures)*, CEB Bulletin d'Information No. 117-F, December 1976.
35. Crandall, S. H., and Mark, W. D., *Random Vibrations in Mechanical Systems*, Academic Press, New York, 1963.
36. Culver, C. G., Lew, H. S., Hart, G. C., and Pirkham, C. W., *Natural Hazards Evaluation of Existing Buildings*, National Bureau of Standards, Building Science Series No. 61, January 1975.
37. Davis, R., Buchanan, B., and Shortliffe, E., Production rules as a representation for a knowledge-based consultation program, *Artificial Intelligence 8*, Elsevier North-Holland Pub. Co., Amsterdam, 1977.
38. Dempster, A. P., Upper and lower probabilities induced by a multivalued mapping, *Annals of Mathematical Statistics,* **38,** 1967, 325–339.
39. Distefano, N., and Pena-Pardo, B., System identification of frames under seismic loads, in: *ASCE National Structural Engineering Meeting*, New Orleans, April 1975.
40. Distefano, N., Preliminary results on the identification problem in non-linear structural dynamics, in: *Proceedings of the Third UCEER Conference*, Ann Arbor, Michigan, May 1974.
41. Distefano, N., Some numerical aspects in the identification of a class of nonlinear viscoelastic materials, *ZAMN 52*, 1972, 389.
42. Distefano, N., and Rath, A., *Modeling and identification in nonlinear structural dynamics*, Report EERC, University of California, Berkeley, 1974.
43. Distefano, N., and Todeschini, R., Modeling, identification and prediction of a class of nonlinear viscoelastic materials, *Int. J. Solids Struct.*, **I,** 9, 1974, 805–818.
44. Distefano, N., and Todeschini, R., Modeling, identification and prediction of a class of nonlinear viscoelastic materials, *Int. J. Solids Struct.,* **II,** 9, 1974, 1431–1438.
45. Dubois, D., and Prade, H., *Fuzzy Sets and Systems: Theory and Applications*, Academic Press, New York, 1980.
46. Duda, R. O., Hart, P., and Nilsson, N. J., Subjective Bayesian methods for rule-based inference systems, in: *Proc. Nat. Computer Conf.*, New York, NY, 7–10 June, 1976.
47. Eykhoff, P., *System Identification-Parameter and State Estimation*, John Wiley & Sons, New York, 1974.
48. Feigenbaum, E. A., The art of artificial intelligence. I. Themes and case studies of knowledge engineering, Invited Paper in: *Proceedings, International Joint Conference on Artificial Intelligence*, Carnegie-Mellon University, Pittsburgh, PA, 1977, 1014–1029.
49. Fenves, S. J., Private communication, 21 October 1977.
50. FitzSimons, N., Techniques for investigating structural reliability, in: *Prob-*

abilistic Methods in Structural Engineering, Edited by M. Shinozuka and J. T. P. Yao, ASCE, 1981, 399–404.

51. FitzSimons, N., and Longinow, A., Guidance for load tests of buildings, *Journal of the Structural Division, ASCE,* **101,** No. ST7, July 1975, 1367–1390.
52. Foutch, D. A., Housner, G. W., and Jennings, P. C., *Dynamic responses of six multistory buildings during the San Fernando earthquake*, Report No. EERL 75-02, California Institute of Technology, Pasadena, CA, 1975.
53. Freudenthal, A. M., Safety of structures, *Transactions American Society of Civil Engineers,* **112,** 1947, 125–180.
54. Freudenthal, A. M., Garrelts, J. M., and Shinozuka, M., The analysis of structural safety, *Journal of the Structural Division, ASCE,* **92,** No. ST1, February 1966, 267–325.
55. Freudenthal, A. M., Shinozuka, M., Konishi, I., and Kanazawa, T., (Eds), *Reliability Approach in Structural Engineering*, Maruzen Co., Ltd., Tokyo, Japan, 1975.
56. Fu, K. S., *Syntactic Methods in Pattern Recognition*, Academic Press, New York, 1974.
57. Fu, K. S., Recent developments in pattern recognition, *IEEE Transactions on Computers,* **C-29,** October 1980, 845–854.
58. Fu, K. S., Ishizuka, M., and Yao, J. T. P., Application of fuzzy sets in earthquake engineering, in: *Recent Developments in Fuzzy Sets and Possibility Theory*, Edited by R. R. Yager, Pergamon Press, New York, 1981.
59. Fu, K. S., and Yao, J. T. P., Pattern recognition and damage assessment, in: *Proceedings Third ASCE EMD Specialty Conference*, University of Texas, Austin, 17–19 September 1979, 344–347.
60. Fukunaga, K., *Introduction to Statistical Pattern Recognition*, Academic Press, New York, 1972.
61. Galambos, T. V., and Mayes, R. I., *Dynamic test of a R/C building*, Dept. of Civil Engineering and Applied Science, Washington University, St. Louis, Missouri, June 1978.
62. Galambos, T. V., and Yao, J. T. P., *Proceedings of Second International Workshop on Code Formats*, Mexico City, 3–5 January 1976, 24–26.
63. Gersch, W., Parameter identification: stochastic process techniques, *Shock and Vibration Digest*, 1975.
64. Guedelhoefer, O. C., Methods for strength evaluation of distressed structures, in: *Probabilistic Methods in Structural Engineering*, Edited by M. Shinozuka and J. T. P. Yao, ASCE, 1981, 327–343.
65. Hanson, J. M., Private communication, 11 June 1977.
66. Hart, G. C., *Estimation of Structural Damage*, J. H. Wiggins Company, Los Angeles, CA, 1976.
67. Hart, G. C., DelTosto, R., and Englekirk, R. E., Damage evaluation using reliability indices, in: *Probabilistic Methods in Structural Engineering*, Edited by M. Shinozuka and J. T. P. Yao, ASCE, 1981, 344–357.
68. Hart, G. C., Rojahn, C., and Yao, J. T. P. (Eds), *Proceedings workshop on interpretations of strong-motion earthquake records obtained in and/or near buildings*, UCLA Report No. 8015, April 1980.
69. Hart, G. C., and Yao, J. T. P., System identification in structural dynamics, *Journal of the Engineering Mechanics Division, ASCE,* **103,** No. EM6, December 1977, 1089–1104.
70. Hasselman, T. K., and Wiggins, J. H., Earthquake damage to high-rise buildings as a function of interstory drift, Private communication, 1982.
71. Healey, T. J., and Sozen, M. T., Experimental study of the dynamic response

of a ten-story reinforced concrete frame with a tall first story, *Structural Research Series*, No. 450, Department of Civil Engineering, University of Illinois, Urbana, IL, August 1978.

72. Hidalgo, P., and Clough, R. W., *Earthquake simulator study of a reinforced concrete frame*, Report No. EERC 74–13, Earthquake Engineering Research Center, University of California, Berkeley, California, December 1974.
73. Housner, G. W., and Jennings, P. C., *Earthquake design criteria for structures*, Report No. EERC 77-06, California Institute of Technology, Pasadena, CA, November 1977.
74. Hsu, D. S., Risk analysis of structures in earthquake engineering, Ph.D. Thesis, School of Civil Engineering, Purdue University, W. Lafayette, IN, August 1977.
75. Hsu, D. S., Gaunt, J. T., and Yao, J. T. P., Structural damage and risk in earthquake engineering, in: *Proceedings International Symposium on Earthquake Structural Engineering,* **2,** University of Missouri, Rolla, MO, 19–21 August 1976, 843–856.
76. Hudson, D. E., Dynamic tests of full-scale structures, *Journal of the Engineering Mechanics Division, ASCE,* **103,** No. EM6, December 1977, 1141–1157.
77. Ibanez, P., Methods for the identification of dynamic parameters of mathematical structural models from experimental data, *Journal of Nuclear Engineering and Design,* **27,** 1974, 209–219.
78. Ibanez, P., *et al., Review of analytical and experimental techniques for improving structural dynamic models*, Bulletin 249, Welding Research Council, New York, June 1979, 44 pages.
79. *Instructions for Classification of Damage Level and Usability of Earthquake Damaged Structures*, Institute of Earthquake Engineering and Engineering Seismology, University 'Kiril i Methodij'—Skopje, Yugoslavia, 1979.
80. Ishizuka, M., Fu, K. S., and Yao, J. T. P., *Inference procedure with uncertainty for problem reduction method*, Technical Report CE-STR-81-24, School of Civil Engineering, Purdue University, W. Lafayette, IN, August 1981.
81. Ishizuka, M., Fu, K. S., and Yao, J. T. P., Rule-based inference with fuzzy set for structural damage assessment, in: *Approximate Reasoning in Decision Analysis*, Edited by M. M. Gupta and E. Sanchez, Elsevier, North Holland Pub. Co., Amsterdam, 1982, 261–268.
82. Kasiraj, I., Low-cycle fatigue damage in structures subjected to earthquake excitation, Ph.D. Thesis, Department of Civil Engineering, The University of New Mexico, Albuquerque, NM, June 1968.
83. Kasiraj, I., and Yao, J. T. P., Fatigue damage in seismic structures, *Journal of the Structural Division, ASCE,* **95,** No. ST8, August 1969, 1673–1692.
84. Kaufmann, A., *Introduction to the Theory of Fuzzy Subsets*, Translated by D. L. Swanson, Academic Press, New York, 1975.
85. Kudder, R., Private communication, 20 April 1977.
86. Kustu, O., Miller, D. D., and Brokken, S. T., *Development of damage functions for high-rise building components*, Report No. JAB-10145-2, URS/John Blume & Assoc., San Francisco, CA, October 1982.
87. *Learning from Earthquakes, 1977 Planning and Field Guides*, Earthquake Engineering Research Institute, Berkeley, CA, 1977.
88. Lee, L. T., and Collins, J. D., Engineering risk management for structures, *Journal of the Structural Division, ASCE,* **103,** No. ST9, September 1977, 1739–1756.
89. Lin, Y. K., *Probabilistic Theory of Structural Dynamics*, McGraw-Hill, New York, 1967.

90. Liu, S. C. and Yao, J. T. P., Structural identification concept, *Journal of the Structural Division, ASCE,* **104,** No. ST12, December 1978, 1845–1858.
91. Luft, R. W., and Whitman, R. V., Earthquake resistance of a historical building, in: *Probabilistic Methods in Structural Engineering*, Edited by M. Shinozuka and J. T. P. Yao, ASCE, 308–326.
92. Marmarelis, P-Z., and Udwadia, F. E., The identification of building structural systems—II. The nonlinear case, *Bulletin of the Seismological Society of America,* **66,** No. 1, February 1976, 153–171.
93. Marsi, S. F., and Anderson, J. C., *Identification/modeling studies of nonlinear multidegree systems, Vol. 3, Analytical and Experimental Studies of Non-Linear System Modeling*, Progress Report AT (49-24-0262), U.S. Nuclear Regulatory Commission, April, 1980.
94. Masri, S. F., Bekey, G. A., Sassi, H., and Caughey, T. K., *Non-Parametric Identification of a Class of Nonlinear Multidegree Dynamic Systems*, University of Southern California, Los Angeles, CA 9007.
95. Matzen, V. C., and McNiven, H. D., *Investigation of the inelastic characteristics of a single story steel structure using systems identification and shaking table experiments*, Report No. EERC 76-20, Earthquake Engineering Research Center, University of California, Berkeley, California, August 1976.
96. Mau, S. T. and Liu, W., Natural periods of buildings in Taipei City, *Journal of Civil and Hydraulic Engineering,* **3,** No. 3, November 1976, 1–8 (in Chinese).
97. Mau, S. T., and Tseng, T. M., Ambient vibration study of a building–foundation system, *Proceedings National Science Council,* **2,** October 1978, 399–406.
98. Meyer, C., Arzoumanidis, S., and Shinozuka, M., Earthquake reliability of reinforced concrete buildings, in: *Probabilistic Methods in Structural Engineering*, Edited by M. Shinozuka and J. T. P. Yao, ASCE, 1981, 378–398.
99. Morrison, D. G., and Sozen, M. A., Response of reinforced concrete plate–column connections to dynamic and static horizontal loads, *Structural Research Series*, No. 490, Department of Civil Engineering, University of Illinois, Urbana, IL, April 1981.
100. Newmark, N. M., and Rosenblueth, E., *Fundamentals of Earthquake Engineering*, Prentice Hall, New York, 1971, 585–587.
101. Okada, T., and Bresler, B., Seismic safety of existing low-rise reinforced concrete building, in: *Developing Methodologies for Evaluating the Earthquake Safety of Existing Buildings*, Report No. UCB/EERC-77/06, Earthquake Engineering Research Center, University of California at Berkeley, February 1977, 51–113.
102. Oliveira, C. S., *Seismic risk analysis for a site and a metropolitan area*, Report No. EERC-75-3, Earthquake Engineering Research Center, University of California, Berkeley, CA, August 1975.
103. Paez, T. L., Random vibration of elasto-plastic structures, Ph.D. Thesis, School of Civil Engineering, Purdue University, W. Lafayette, IN, December 1973.
104. Ravindra, M. K., and Galambos, T. V., Load and resistance factor design for steel, *Journal of the Structural Division, ASCE,* **104,** No. ST9, September 1978, 1337–1352.
105. Rodeman, R., Estimation of structural dynamic model parameters, Ph.D. Thesis, School of Civil Engineering, Purdue University, W. Lafayette, IN, August 1974.
106. Rodeman, R., and Yao, J. T. P., *Structural identification—literature review*,

Technical Report No. CE-STE-73-3, School of Civil Engineering, Purdue University, W. Lafayette, IN, December 1973, 36 pages.

107. Rosenblueth, E., and Yao, J. T. P., On seismic damage and structural reliability, Unpublished Technical Note, November 1977.
108. Rudd, J. L., Yang, J. N., Manning, S. D., and Yee, B. G. W., Damage assessment of mechanically fastened joints in the small crack size range, in: *Proceedings Ninth U.S. Congress of Applied Mechanics*, ASME, 1982, 329–338.
109. Sage, A. P., and Melsa, J. L., *System Identification*, Academic Press, New York, 1971.
110. Schiff, A. J., Identification of large structures using data from ambient and low level excitation, in: *System Identification of Vibrating Structures*, Edited by W. D. Pilkey and R. Cohen, ASME, 1972, 87–120.
111. Scholl, R. E., Kustu, O., Perry, C. L., and Zanetti, J. M., *Seismic damage assessment for high-rise building*, Report No. URS/JAB-8020, URS/John Blume & Assoc., San Francisco, CA, July 1982.
112. Seed, H. B., Idriss, I. M., and Dezfulian, H., *Relationships between soil conditions and building damage in the Caracas earthquake of July 29, 1967*, Report No. EERC 70-2, Earthquake Engineering Research Center, University of California at Berkeley, February 1970.
113. Shafer, G., *A Mathematical Theory of Evidence*, Princeton University Press, Princeton, NJ, 1976.
114. Shibata, H., *A comment on fuzzy sets. Summary of papers on general fuzzy problems*, Report No. 2, Tokyo Institute of Technology, Tokyo, Japan, December 1976, 101–106.
115. Shinozuka, M., and Kawakami, H., *Underground pipe damage and ground characteristics*, Technical Report No. CU-1, Department of Civil Engineering and Engineering Mechanics, Columbia University, New York, June 1977.
116. Shinozuka, M., and Yang, J. N., Optimum structural design based on reliability and proof-land test, Proceedings of the Reliability and Maintainability Conference, *Annals of Assurance Science*, **8,** July 1969, 375–391.
117. Shortliffe, E. H., *Computer-Based Medical Consultations: MYCIN*, American Elsevier, 1976.
118. Shortliffe, E. H. and Buchanan, B. G., A model of inexact reasoning in medicine, *Mathematical Biosciences,* **23,** 1975, 351–379.
119. Shortliffe, E. H., Buchanan, B. G. and Feigenbaum, E. A., Knowledge engineering for medical decision making: a review of computer-based clinical decision aids, *Proc. IEEE,* **67,** Sept 1979, 1207–1224.
120. Sozen, M. A., Review of earthquake of R/C buildings with a view to drift control, in: *State-of-the-Art in Earthquake Engineering 1981*, Edited by O. Ergunay and M. Erdik, Turkish National Committee on Earthquake Engineering, Ankara, Turkey, 1981, 383–418.
121. Srinivasan, M. G., Kot, C. A., Hsieh, B. J., and Chung, H. H., *Feasibility of dynamic testing of as-built nuclear power plant structures: an interim evaluation*, Report No. NUREG/CR-1937, ANL-CT-81-5, U.S. Nuclear Regulatory Commission, Washington, D.C., May 1981.
122. Sues, R. H., Wen, Y. K., and Ang, A. H-S., Safety evaluation of structures to earthquakes, in: *Probabilistic Methods in Structural Engineering*, Edited by M. Shinozuka and J. T. P. Yao, ASCE, 1981, 358–377.
123. Tang, J.-P., Random fatigue in earthquake engineering, Ph.D. Thesis, Department of Civil Engineering, The University of New Mexico, Albuquerque, NM, August 1971.

124. Tang, J.-P., and Yao, J. T. P., Expected fatigue damage of seismic structures, *Journal of the Engineering Mechanics Division, ASCE,* **98,** No. EM3, June 1972, 695–709.
125. Ting, E. C., Chen, S. J. Hong, S. T., and Yao, J. T. P., *System identification damage assessment and reliability evaluation of structures*, Technical Report No. CE-STR-78-1, School of Civil Engineering, Purdue University, W. Lafayette, IN, February 1978, 62 pages.
126. Toussi, S., System identification methods for the evaluation of structural damage. Ph.D. Thesis, School of Civil Engineering, Purdue University, W. Lafayette, IN, August 1982.
127. Toussi, S., and Yao J. T. P., *Identification of hysteretic behavior for existing structures*, Technical Report No. CE-STR-80-19, School of Civil Engineering, Purdue University, W. Lafayette, IN, December 1980.
128. Toussi, S., and Yao, J. T. P., *Hysteresis identification of multi-story building*, Technical Report No. CE-STR-81-15, School of Civil Engineering, Purdue University, W. Lafayette, IN, May 1981.
129. Toussi, S., and Yao, J. T. P., Assessment of structural damage using the theory of evidence, *Structural Safety,* **1,** 1982/1983, 107–121.
130. Udawadia, F. E., and Kuo, C. P., *Nonparametric identification of a class of nonlinear close-coupled dynamic systems*, Report University of Southern California, Los Angeles, CA.
131. Udawadia, F. E., and Marmarelis, P. Z., The identification of building structural systems—I. The linear case, *Bulletin of the Seismological Society of America,* **66,** No. 1, February 1976, 125–151.
132. Wang, T. Y., Bertero, V. V., and Popov, E. P., *Hysteretic behavior of reinforced concrete framed walls*, Report No. EERC 75-23, Earthquake Engineering Research Center, University of California, Berkeley, California, December 1975.
133. Wiggins, J. H., Jr., and Moran, D. V., *Earthquake Safety in the City of Long Beach Based on the Concept of Balanced Risk*, J. H. Wiggins Company, Redondo Beach, California, September 1971.
134. Whitman, R. V., Reed, J. W., and Hong, S. T., Earthquake damage probability matrices, in: *Proceedings 5th World Conference on Earthquake Engineering*, Rome, Italy, 1973.
135. Whitman, R. V., *et al.*, Seismic resistance of existing buildings, *Journal of the Structural Division, ASCE,* **106,** July 1980, 1573–1592.
136. Yang, J. N., and Shinozuka, M., On the first-excursion probability in stationary narrow-band random vibration, I, *Journal of Applied Mechanics, ASME,* **38,** No. 4, December 1971, 1017–1022.
137. Yang, J. N., and Shinozuka, M., On the first-excursion probability in stationary narrow-band random vibration, II, *Journal of Applied Mechanics, ASME,* **39,** No. 3, September 1972, 733–737.
138. Yang, J. N., and Trapp, W. J., Reliability analysis of aircraft structures under random loading and periodic inspection, *AIAA Journal,* **12,** No. 12, December 1974, 1623–1630.
139. Yao, J. T. P., Assessment of seismic damage in existing structures, in: *Proceedings U.S.–S.E. Asia Symposium on Engineering for Natural Hazard Protection*, Edited by A. H-S. Ang, Manila, Philippines, 388–399.
140. Yao, J. T. P., Damage assessment and reliability evaluation of existing structures, *Engineering Structures,* **1,** October 1979, 245–251.
141. Yao, J. T. P., Damage assessment of existing structures, *Journal of the Engineering Mechanics Division, ASCE,* **106,** No. EM4, August 1980, 785–799.

142. Yao, J. T. P., Identification and control of structural damage, *Solid Mechanics Archives,* **5,** No. 3, August 1980, 325–345.
143. Yao, J. T. P., and Munse, W. H., *Low-cycle axial fatigue behavior of mild steel*, ASTM Special Technical Publication, No. 338, 1962, 5–24.
144. Yao, J. T. P., Toussi, S., and Sozen, M. A., Damage assessment from dynamic response measurements, in: *Proceedings Ninth U.S. National Conference on Applied Mechanics*, ASME, June 1982, 315–322.
145. Zadeh, L. A., Fuzzy sets, *Information and Control,* **8,** 1965, 338–353.
146. Zadeh, L. A., Outline of a new approach to the analysis of complex systems and decision processes, *IEEE Transactions on Systems, Man and Cybernetics,* **SMC-3,** No. 1, January 1973, 28–44.
147. Zadeh, L. A., Fu, K. S., Tanaka, K., and Shimara, J., *Fuzzy Sets and Their Application—Cognitive and Decision Processes*, Academic Press, New York, 1975.

Index